Rolling Through the Isles

Rolling Through the Isles

A Journey Back Down the Roads that Led to Jupiter

TED SIMON

Little, Brown

LITTLE, BROWN

First published in Great Britain in 2012 by Little, Brown

Extract from the Sloane Grammar School website reproduced
by kind permission of Mark Foulsham

A CIP catalogue record for this book
is available from the British Library.

ISBN 978-1-4087-0218-5

Typeset in Baskerville by M Rules
Printed and bound in Great Britain by
Clays Ltd, St Ives plc

Papers used by Little, Brown are from well-managed forests
and other responsible sources.

MIX
Paper from
responsible sources
FSC® C104740

Little, Brown
An imprint of
Little, Brown Book Group
100 Victoria Embankment
London EC4Y 0DY

An Hachette UK Company
www.hachette.co.uk

www.littlebrown.co.uk

For Meg and Eliot

1

A Bump in the Road

Because of my reputation as a motorcyclist a magazine editor asked me, a few years ago, where I would take my last ride and I was rather taken aback. I don't spend much time thinking about the end of things, since age seems to have left me fairly intact so far, but I suppose it was a reasonable request, given that I was already well into my seventies.

Of course, his request begs the question: how would I know that it was my last? Well, there'd be one way of making sure. I could leave a neat pile of personal effects, topped by a helmet, at the edge of a very high cliff. I've known at least one person who took that final ride, and I can imagine that, as a way of pre-empting a painful end, it might be thrilling, but I hope I can think of better ways to hang up my helmet.

So, eschewing suicide, I said I'd plump for nostalgia. There remain many places on earth – China, Japan, Russia, to name a few – where I would still like to ride a bike, but for the ulti-mate excursion I said I thought I'd circle back to the beginning, and try to recover some of those early experiences in the land where I was raised. And so, having thought of it, it became almost inevitable that I would do it.

Since I moved away from England in 1968 I've seen rather little of the British Isles, but when I was at school, a few years after the end of the war (do I really have to tell you which one?), I used to hitch-hike all over the place, sometimes alone, sometimes with a school friend or two. It was the only way we could afford to get around.

There was a girl in Glasgow I was smitten with and I used to boast that I could get there from London's North Circular Road in eighteen hours. In the forties that was pretty good time. It went like clockwork, and my lifts were always in lorries, because in those days cars were still a rarity. Towards evening you'd get a bus out along the Finchley Road to a junction on the North Circular and catch a lorry on its way north up the A6. The drivers liked the company, it helped them to stay awake, so getting a lift was pretty easy. Usually they were headed for Liverpool or St Helens – I was never lucky enough to find one going all the way to Scotland – which meant a change at Warrington.

These were not big lorries by modern standards, though they seemed big to me then. Scammell, Dennis, Foden, Bedford: some of those names have since disappeared, and I don't remember too much about them. I wasn't as curious as I should have been. The cabs were primitive, scarcely insu-lated and usually heated by the engine block, which sat between me and the driver, with a greasy old horse blanket thrown across it. They might have carried anything from six to ten tons of cargo under a canvas cover. Going north, I never knew what was in the back, but coming south I can remember the scent of coal and, on one occasion coming down the A1, fish.

The drivers were usually interesting characters, and very different from people I met otherwise. One I recall spoke only eight words in as many hours on the road to Glasgow. Putting his hand over the engine he said, 'This thing's as 'ot as a fuckin' whorehouse.'

On a clear starlit night, lorry drivers would sometimes switch off all their lights and, remarkably, after a moment of adjustment the whole landscape would open up and you could see so much better. Of course this was illegal. They were always on the lookout for police, in their low-slung black Vauxhalls, who, I was told, crept up on them and clung to their tails where they couldn't be seen, waiting for them to do something naughty so that, in an age before sirens, they could ring their bells. I recall at least one story of the Lorry Driver's Revenge. Sharp and sudden braking could have a gratifying crumpling effect on the sharp nose of a Vauxhall, causing much constabulary contumely.

No doubt the police had other reasons to be interested in what lorry drivers were up to. In those days the black market was flourishing and I can't imagine that some of them didn't have a hand in provisioning it, but I was an innocent and never thought to ask about such matters, which was probably just as well.

So, the lorries I travelled on would get to Warrington in the middle of the night and turn off west to St Helens, leaving me to spend some hours near the municipal rubbish dump waiting for dawn. Then I'd catch another going up over Shap through Carlisle to Glasgow. If I was lucky he'd stop at the Jungle Café, an honourable and much-loved institution where you could get a huge breakfast. Bacon, eggs, mushrooms, tomatoes, beans, lashings of toast and sweet tea ... In those days of rationing that was a fabulous treat.

I remember going up the A1 and wandering, wide-eyed and frozen, around Edinburgh and the Aberdeen fish market. I hitched to the West Country too, where the sun always seemed to be shining. I felt very much at home in those old market towns of Somerset, Dorset and Devon, and gloried in the warmth of the light reflected off old stone. Over the years, whenever I was asked about my favourite part of England, I always thought of Dorset and points west, although in truth this had more to do with my ignorance of most of the rest of the country.

That, I thought, could be my last ride: to rediscover the United Kingdom before the age of motorways, very slowly, between towering hedges, over moors, along leafy lanes and back roads, to places I've been, and even more I haven't. There would be all of Ireland to conquer, and much of the north. There would be pubs, inns and B&Bs galore. I'd be on a small bike, very understated, no fancy gear or space-age suit, prefer-ably just carrying a toothbrush and a credit card; I'd never do more than a hundred miles a day.

Perhaps I'd find Trench Hall, the stately home where I spent two years as an evacuee, and I could reminisce about the house where my sweetheart lived on the Dumbarton road.

And then, as I thought about it, more and more episodes of

my early life came to mind, some hilarious, some sombre. There was the distinctly odd year I spent on a provincial paper, in and around Barrow-in-Furness, when I was awaiting Her Majesty's pleasure, with trips to Blackpool and the Lake District. And there were even stranger occurrences when I was finally recruited to the ranks of the RAF: I managed to turn my National Service into a working holiday, with the help of stars like Peter Sellers and Spike Milligan.

So, if I did eventually arrive at Beachy Head perhaps I'd push my bike over the edge, walk away . . . and write a book.

Certainly the idea appealed to me. I thought I could make the trip during the late summer of 2009, when the weather was most likely to be fine, taking my time, then write the book, nothing too voluminous, through the winter for my seventy-ninth birthday in May 2010.

The idea was enthusiastically received so I planned my season in Europe. One might think that it was a rather ambitious plan for a seventy-eight-year-old, but it didn't seem so at the time. The journey from San Francisco is so long and expensive that I always try to make as much of my visits to Europe as I can. The mammoth schedule was only possible, financially, if I travelled around on a bike, and a friend in Germany keeps my small BMW in his garage just for that purpose. So, as I usually do, I flew from San Francisco to Frankfurt to collect it and then, rather like a migrating bird, began the long and comforting circuit of old friends and familiar places, thinking all the while how extraordinarily fortunate I was to be free to travel like this, to have such warm relationships with so many people, and still to be fit enough to do it and enjoy it.

I went first to Stuttgart where Dunja, an ex-girlfriend, who was now a dentist, fixed a bothersome tooth. The next night I spent with Roland, a Bavarian industrialist who is also my neighbour in California – he had fallen in love with the primitive beauty of the Wild West many years ago. He happened to be at his home near Munich and wanted me to see the factory where he had made his fortune – a marvel of Teutonic precision that produces, among many other things, extrusion nozzles for the plastics business. From there I crossed over Switzerland to Voiron, where I renewed my allegiance to a fierce and complex French liqueur called Chartreuse that comes from a monastery in the area, and which I had learned to love in Paris in the fifties.

The following day I was in Montpellier with other old friends from the days when I lived in the South of France, and it was there, a few days later, that I met my two favourite Spaniards, also travelling by bike, who publish my books in Spain. So, with Angel and Teresa, I rode on a leisurely three-day trip to Madrid where we did a book-launching party and took in a *fiesta* for good measure. Teresa's father is a butcher, so I was able to carry away a leg of cured ham, the highly prized one with a black foot, though why *patas negras* should make better ham I don't know. He cut it up and wrapped it in portions that would fit in my panniers, and the bone he chopped up for soup. In Toledo, that ominous citadel of stone and steel, I bought a long, thin carving knife like a sword, so from then on I shed slices of ham wherever I went.

North from Madrid I went to see the famous museum in Bilbao, which looks as though a pile of aluminium leaves have blown together to make a building. Going back towards France, I spent some time with a friend from my earliest days in Fleet Street, who has set up home on the Pilgrim Route to Compostela. And then, leaving Spain for France on the *autoroute*, several of the spokes on my rear wheel suddenly collapsed, the tyre went *phut*, and I was lucky to get off the road in one piece. A wry but friendly Frenchman hauled me off to his shop in Biarritz, where it turned out I was three spokes short of the necessary number. Because of a silly combination of weekend and national holiday, I was condemned to spend a week there (a penalty many would be happy to pay) before new spokes arrived from Munich and I could finally make my way up France.

By the time I got to Trôo, a troglodytic town on the Loire and yet another place where I would love to live, the ham bones had suffered and were beginning to look as though I'd dug them up. Kate, a friend from San Francisco, who has a cave there, was eagerly awaiting them, and I couldn't bear to disappoint her. Since we were, so to speak, in the stone age anyway, I made an executive chef's decision. We scraped off the moss and made enough soup for the village. Everyone was happy, and nobody died.

By now it was June and time to cross over to England. David Wyndham in Dorchester had asked me if I would show my film at his dealership in favour of the wounded soldiers coming back from Afghanistan and I was glad to do so (much as I wished they had never been sent to Afghanistan in the first place).

Of course, as a courtesy to my friends, I had let them all know what I had in mind to do later that summer, and Dave was keen for me to use one of the smaller bikes he had in his shop, but they didn't fit the rather vague image I had in mind.

Among my friends, of course, was Stephen Burgess, who had lent me the BMW I had ridden around the world in 2001. He has always insisted that I could use it any time, but I was quite sure that such a big bike wouldn't do at all for my slow saunter down Memory Lane.

Trying to identify what this bike would be, I began to wonder whether I had trapped myself with an idea that couldn't be realised. And then, while I was in London, a strange little machine came tootling past me on Putney High Street and I was sold.

It was a scooterish affair with two wheels at the front. I had never known that such a thing existed. I looked it up on the Internet and there it was, the MP3, half scooter, half bike, made by Piaggio who also made the Vespa. It was strange enough, I thought, to unseal the lips of all the bibulous bystanders in Blighty. I had in my mind's eye an enticing, if rather quaint, picture of myself as an elegant gentleman of a certain age, probably in kid gloves and a silk scarf, rolling up to a rustic coaching inn where loquacious locals lounging around with foaming pints would be intrigued by this phenomenon – 'Why, sir, may we enquire what brings a fine gentleman like yourself to these parts, and on such an unusual conveyance?' – thus unleashing a flood of tales and reminiscences, which I would dutifully transcribe that night behind mullioned windows by the light of a guttering candle ... Well, maybe the picture was somewhat overdrawn, but you get the idea.

After the show in Dorchester I rode over to visit Stephen and to tell him what I had in mind. It was purely to satisfy his natural curiosity. Steve is a classic biker, in the pure BMW mould, and I didn't imagine that he would approve. In fact, he gazed at me from the depths of his armchair as though he had encountered some strange new species of lizard, but then said, in mock-weary terms, 'Well, I suppose that's where I come in.'

When I realised he felt obliged to provide me with this monstrosity, I was quite embarrassed and insisted that I had never had such an idea, and was quite capable of buying one for myself, thank you, but he said, 'The trouble with you is, you can't take yes for an answer,' and promptly disappeared into the next room. I waited uneasily, wondering if he had gone to fortify himself against the shock.

Well, he came back in minutes, quite matter-of-fact – might have gone for a pee – and said, 'That's done, then. I found one on eBay. It's not new. I thought, given the way you treat your bikes, that a shiny new one would be a bit of a waste. It belonged to a professional gambler, which seems suitable.'

'You've *bought it!*' I said, in hushed tones.

'Yep. Wally can fetch it, get it down to Dorchester and look it over.'

So that was that. To refuse such generosity now, I thought, would be an insult. It's not what I intended but there is an art to receiving gifts as well as to giving them, and I learned it well on my travels.

There were still more things I wanted to do, in England and Europe, before I began my adventure. I took my bike to a meeting near Derby, and went on to another in Wales. I met my publishers in London, and crossed the Channel to Normandy to visit lovely Cathérine Germillac, who had once taken her Yamaha halfway round the world and was now recovering from a difficult illness. From there I crossed France to visit Bruno

Bouvery, my erstwhile companion in South America thirty-five years ago, and then went back to Duisburg to make sure the bike was OK. When Dirk Erker, my good friend, mechanic and motorcycle nanny, confirmed that everything was '*in ordnung*' and told me what an idiot I was, I was finally free to ride through Poland to Ukraine to spend a week with Lida, whom I'd met on my walk through Eastern Europe in 1993.

So, in a matter of six weeks I had revisited seven different periods of my life: my first priceless years in Paris, my life as a newspaper editor, my 180-degree turn to the South of France, my ride around the world, my move to California, my walk from Germany to Romania, and my return journey around the world. I was feeling extremely fortunate, and looking forward to going even further back, into my childhood, in England. And then, as luck would have it, just after leaving Lida and riding at a leisurely pace to the Polish border, I hit the road, head first.

It was a silly accident and, like all my accidents, it was both good and bad, but it was also life-changing. I woke up at the roadside, having enjoyed a few minutes of unconsciousness, without the slightest idea of what had happened. Another man on a smaller bike had apparently been involved and claimed I had come up behind him at speed and hit him. I knew this was virtually impossible, but it took me quite a while to deduce that he must have turned left in front of me from the gutter – where most Ukrainians ride what few bikes there are – without looking or signalling. However, I was more concerned with the broken bone in my left arm.

Lida arrived at the scene of the crime, did an extraordinary job of retrieving all my stuff, and delivered me to the tough love of the Ukrainian medical establishment, which drove me around from hospital to hospital over some very rough roads before eventually releasing me to her care. I resigned myself to convalescence in a quiet Ukrainian village not far from L'viv. My only live contact with the outside world was a visit from Georg, a German friend whom I first met in Penang in 1975 (back we go again!). Because he was a doctor he showed interest in my concussion and said I should probably have my head looked at some time although, he thought, it was probably all right.

There was nothing to be done but wait for the arm to heal and sort out the bumps on the bike so it was six weeks before I could think of leaving. The arm had mended and I felt more or less all right. By then Lida and I had become much more than friends, so that was the good part of the accident. The worst was still to come. I rode the bike back to Germany in a couple of days, blissfully unaware that the forks were twisted (not that this was anything new: I've ridden other bikes in the same condition for considerable distances. I could even claim it as a speciality of mine). But Dirk Erker was not impressed. He said I couldn't go on riding the bike until it was fixed, and if I was determined to go on with my enterprise I would have to fly to London.

This turned out to be a mistake. After a difficult flight, I got into London much the worse for wear and arrived at a friend's doorstep in an embarrassing state of incontinence. After a couple of days of uncertainty, I took Georg's advice and went to the Chelsea and Westminster Hospital to 'have it looked at'. What they saw when they looked at the CT scan had me in bed within minutes and under twenty-four-hour surveillance. So, my tour of the British Isles began in a close encounter with the

National Health Service, and although I wish it hadn't had to happen that way, I'm grateful, really, for the experience.

I've been in hospital five times in my life and survived them all, for which I give thanks. Once, in Kenya, they had to mend a broken leg, but what's remarkable about the other four times is that – quite correctly – they did absolutely nothing, and three of those hospitals were in England. As a kid I had an obscure glandular swelling and they just watched it go away. As an adolescent they put me in a nice private bed until I'd recovered from polio, which was really all they could do. And on this occasion they watched and waited to see whether the subdural haematoma they had discovered in my head was going to get better or whether they would have to drill through my skull. I'm very glad they waited. On the whole, I think American doctors interfere too much. They prescribe more than they should, and cut when they shouldn't, and I hope I never have to eat these words.

It seems that when you have one of those subdural things you shouldn't go flying. The change in pressure doesn't do it any good. Apparently there's a place in the middle of the brain that controls certain motor functions and, under pressure, causes you to pee rather freely. When they let me out of the hospital I clung to the idea that I might still do the journey. After all, they had only forbidden me to fly. So I went down to Stephen's place in Hampshire where my peculiar little trike was now tucked away unceremoniously in one of his garden sheds. He was kind enough to let me ride it round the lawn, but it was obvious to everyone else, and last of all to me, that I was in no shape to take to the open road. The whole thing would have to be put off for another year.

Some very kind friends took me in and gave me time to recover. According to them, and others, I was not entirely myself for a couple of months, but I was quite present enough to form an opinion of 'the National Health': I was really impressed. It gives me great pleasure to be able to say this. In America, where they have the most expensive and divisive medical system in the Western world, it is customary to protect the profits of the insurers and the providers by spreading slander about national health systems, and the NHS probably gets more than its fair share. In my experience, the treatment was prompt and efficient, the ward was pleasant, the nurses were sympathetic and effective, the consultants were delightfully free of pomposity and happy to provide information, the food was good – there were menus and choices – there was personal TV, radio and Internet, and there was no financial pressure. All in all it would have been hard to find fault.

2

The Man in the Brown Suit

A year later, on 1 July 2010, fully recovered and newly married, I flew to England to try again. Unfortunately I was forced to leave my brand-new gorgeous wife at Frankfurt railway station, the British government having decided to keep all honest visa applicants out to make room for the dodgy ones slipping in. I took the train alone to the airport where British Airways was waiting to fly me to London.

After the ride to the Airbus, all the passengers and I got in and we drove off to England. This is something that happens at Frankfurt. Instead of taking off, planes simply drive across miles and miles of tarmac. After about ten minutes' driving across the airport I fell into a doze so I don't know which *autobahn* we used or how we got across the Channel, but when I woke up we were still driving, on the last leg to Terminal 5 at Heathrow.

I had never been to the new Terminal 5 before and I was pleased to see, when we got inside, that the architect (I presume there *was* an architect) had restrained himself from employing any fancy new technology for moving people around. He or she had made the sensible decision that, after the long drive from abroad, what people needed was a brisk five-mile walk to the baggage place, thus keeping alive the tradition of all the other Heathrow terminals.

There was no time wasted either because the bags – which, I believe, an army of asylum supplicants brings in from the plane by hand – arrived well after we did, giving us all a chance to get to know each other. Eventually they began winding the carousel around, which must be really hard work.

I rolled my stuff past the yawning Customs officers and was finally shocked out of my dream by the bright red hair of my friend Lois Pryce, who had come to fetch me. Lois is a wonderfully bubbly woman, with a sibilant voice and steel tendons, and her husband, Austin Vince, famous for his overalls, is one of the most refreshingly direct people I know. They have both ridden motorcycles in very unlikely places and written about it. Fortunately for me, they live very conveniently on a canal boat not far from the airport, and they had offered to stuff me into a hole somewhere under the bows.

Lois, for logistical reasons, had also brought along a neighbour of hers, Tim, who plays the double bass and drives one of those old cars with retro-fitted seat belts and doors that crunch. He turns out, appropriately enough, to have played with Bill Haley and the Comets, among many other illustrious musicians, so the rest of the morning was consumed by hilarity. Soon I was off again, to be lunched by my publishers. They kept buying me lunch and I was not ashamed to eat at their table because at the time they were publishing that enormously profitable vampire series by Stephenie Meyer.

Well, my journey to their office on the Embankment started at West Drayton, and the train, which pulled in just after I arrived at the station, was apparently thirty-five minutes late, but I didn't pay much attention to that. I was looking forward to seeing Paddington again. Paddington has always been my favourite station: it used to be the home of the Great Western Railway, and when I went there as a kid it was always the summer holidays, and I was to be transported west (where the sun always shines). So when we arrived I was rather perturbed to see that they seemed to be pulling it down.

When I got off the train, instead of seeing the spacious station I knew, I found myself in an alley of corrugated sheet-metal screens, with occasional glimpses of jagged concrete and twisted rebar between the gaps. I told an obliging woman in some sort of uniform that I wanted to go to Blackfriars and she directed me to keep walking along that corridor until I got to the other end of the station, about a quarter of a mile, and then to go upstairs, which I did. And there, behind a desk, sat an equally obliging man who told me to follow some coloured stripes on the ground back to where I had come from, but by a different route this time, along a proper platform.

A walk never did anybody any harm, and I was glad of this one because it led me past one of the stranger sights of recent memory. Alongside the platform I saw several hundred people in a concentration camp, penned in by plastic barriers. Most of them were young and many carried backpacks. They seemed oddly content with their lot so I asked one what he was doing there. He said he was hoping to go to Swindon, but he didn't know about the others.

Bemused by all this, I went on to Blackfriars station, had a very good lunch, and returned to what was left of Paddington.

This was now a scene of unimaginable chaos. Huge numbers of people were swarming across the concourse, back and forth, in tidal waves, while others stared upwards at screens with their necks locked in position. At last I was given an explanation. It appeared that someone in need of copper had stolen the signal cables from West Drayton, and mine had been the last train in.

This mischievous theft aroused fond memories from my youth. Back in the forties my stepfather, Bill, had worked on the Underground as an engineer and used to tell me stories from that enchanted subterranean world – stories that could make your hair stand on end – like the one about the lift that was discovered to be hanging by the last few strands of a single cable, or the escalator that was found to be running with a complete tread missing, and others that made me collapse with mirth.

At that time non-ferrous metals were at a premium. My dad told me about generators and sub-stations abandoned and forgotten in the labyrinth of tunnels, where canny workers on their way to the pub could pick up enough copper to buy a few pints. The funniest story was about a chap in a brown suit with a briefcase who went down the circular staircase at Notting Hill Gate to the Gentlemen's Convenience. I particularly remember visualising the brown suit. He came back up a little later, soaking wet from head to foot because some likely lad had removed the lead S-bends that used to connect the overhead cisterns to the down pipes. The laughter went on longer than usual as we tried to imagine why he would have pulled the chain while he was still sitting down.

Nothing nearly so hilarious was happening now at Paddington. There was no train back to West Drayton, or indeed to anywhere else, so I had to take several tube trains on an endless diversion to Uxbridge. On the station platforms, as I went, I heard announcements about other signals that were apparently failing in sympathy, like dominoes, stranding people in various parts of the metropolis.

The next day I was in a pub across the canal, watching England being thoroughly thrashed by Germany in the World Cup. All around me Brits were supposedly being traumatised as the goals poured past the unhappy keeper, but there was something odd about the experience. In a curious way the emotions seemed fake, as though it was no more than they had expected. After the game they all went back to their gastropub meals as though nothing much had happened while I, the expatriate, went back to the boat dejected.

All in all, it seemed to me that I had arrived to explore England at an interesting time.

3

An Ingenious Device

The bike was waiting for me down in Dorchester, where the faithful Wally had taken it back to CW Motorcycles. I won't be bashful about trumpeting their virtues. What they have done for me in the past may or may not have helped their reputation, but they did it, I know, for the sheer pleasure of being involved in my global transgressions. I have been in their shop often enough to witness their enthusiasm, and the care and concern they lavish on their customers. In the past I'd found most British mechanics a dour and miserable lot, but CW have changed my view altogether.

All the same, I was quite nervous about confronting them with a scooter, even if it did have two wheels at the front. Would I notice the contempt leaking out of them, like the oil from my old Triumph? Would there be mutterings of 'treason'? I remembered asking Steve, the year before, if he had tried it out and he said, 'Yes, I did. It's terrible.'

But, to my surprise and relief, Wally, who has most to do with the maintenance side of things, was excited about it. 'I think it's marvellous,' he said. 'I've had a go on it. Very clever how they get those wheels to work together.'

It's very compact, almost rectangular, and has a pretty useful amount of space under the seat. My computer case fits in there perfectly, together with other useful items, and I've got a shiny red waterproof bag to tie on behind that holds most of my clothes. And there's another space at the back that's just big enough to take the helmet, so all in all I'm well fitted out.

But for what?

Drifting through my childhood memories at home, I was captivated by images of the different, older England they suggested so vividly. I was seduced into believing that if I just went back in the right frame of mind it would reveal itself. In a hypnotic trance, I had given no thought to the problem of piercing the armour of contemporary society, and now that I was set to go I couldn't see my way. My dream of wandering through the little leafy lanes of England seemed suddenly remote. How would I ever find them, without the local knowledge that would take an age to acquire?

Studying the road maps I saw that, reasonably enough, they were designed to enable a person like me to get from A to B in the most efficient way, which meant, inevitably, guiding me on to the motorways and the multitude of fast dual-carriage A-roads that have laid a tar and concrete lattice over England.

To plot a route that would avoid these tunnels of pounding rubber and roaring exhaust would take for ever and probably be futile. How could I possibly tell, looking at the atlas, which of those thin lines in the cells of the lattice would take me back where I wanted to go?

Of course I had my TomTom. (In England my friends call it a 'satnav'. This is the one and only time you will see that detestable word in this book. With 'infarct' and 'segue', it is one of the great insults to our language.) Perhaps, I thought, the TomTom could help me to find what I was looking for, and I prepared to give it a run. Then I came up against the only thing that has really annoyed me about the MP3: there was no cigar lighter, nowhere to plug it in. How could Piaggio imagine that the dashing man-about-town, like the professional gambler who once owned this very machine, would not want to smoke his cigar between casinos? Like Winston Churchill on the prow of his Thames barge, an accomplished, modish rider could probably blast his way through a box of Havanas between Chelsea and the Haymarket.

Anyway, one of the lads at CW was good enough to rig something up, taping a wire from the battery across the floor of the bike and up to the dashboard, and the following morning I prepared for a short trial run into the glorious July landscape.

But first it struck me that it might be wise to buy some petrol, at which point it also occurred to me that I had never noticed a suitable hole to pour it into. In the car park of the Beggars Knap, a B&B where I was staying, I began my search, casually at first and then with ever-increasing frustration, but nowhere could I find an access point that wasn't screwed down. I examined the black plastic shell of my MP3 in growing desperation. Incomprehensibly, even though the manual discusses, in minute detail, every last fuse and grommet, nowhere does it show where the petrol goes in. It does tell you that a flick of the ignition switch will cause a cover to pop up, and I flicked away, but there was nary a pop. Until at last, resorting to the methods of Sherlock Holmes, I deduced by pure reason where the elusive orifice would have to be, and found that the two-inch tape holding down the cable to my TomTom was also very cleverly disguising and holding down the petrol cap right under my feet.

I know very well that this sort of stuff is always happening to most people and you may wonder why I've wasted your time going on about it, but I see it as yet another milestone on my road to ruin. It seems to me that there was a time when things went my way, and I do mean specifically *things*, such as inanimate matter, objects, bits of stuff. When I threw a log on to a pile, it landed more or less where I threw it and stayed there. Today it bounces off on to my foot. When I drop something on the floor it slithers three yards away down a hole. If I'm ever lucky enough to find it again I have to bend down three times before I get a grip on it. Surely it never used to be like that?

I remember very well the first inkling I had that things were turning against me. I was trying rather urgently to fasten a tarpaulin in a gale, and I couldn't get one of the ropes free because it had wrapped itself around a bottle, just an ordinary empty wine bottle (of which, I must admit, there were many). It's the absurdity, the sheer impossibility of it, that gets to me. Never again in my lifetime, never in anybody's lifetime, will a rope contrive to fasten itself to a bottle. Imagine climbing a mountain and your partner says, 'It's OK. I've got the rope fastened to a bottle.' Or you've reached port in a storm and the skipper says, 'Come on, lad, tie us up to that bottle.'

Well, that was an extreme example, but it often takes a shock to tip you off that things are changing, and since that fateful day I have noticed the trend. Matter no longer co-operates. So far, happily, it is only small things that defy me. In large matters, such as motorcycles, my luck is holding out, but it may only be a matter of time.

Actually, it was one of those small things, an ignition key, that caused me to choose Yeovil as the destination for my first trip. The MP3 has two devilishly intricate ignition keys. One of them, the everyday one, performs many small miracles. Twist it one way and it opens the helmet compartment at the back. Twist it another way, and it opens the petrol cap. Push it in and turn it, and it locks everything. And then it has a little blip on it that unlocks the seat to reveal the luggage space.

The other key, the master key with a built-in computer chip, does most of these things but it doesn't have the blip: without it one has to rummage around blindly through the helmet compartment looking for a latch. Life without the blip would be a nuisance, and somehow, unaccountably (I insist), I managed to lose the key with the blip within twenty-four hours of getting the vehicle.

Now, here is another reason why I love CW. To lose the key to your bike as soon as you get it is the kind of behavioural stupidity that should banish a man from decent company for ever. When it became known at the shop that I had lost it (and I couldn't avoid telling them), I braced myself for sniggers and barely concealed whispers all round. 'He's lost it, all right,' or 'No good getting him another. He'll never find the keyhole.' But not bit of it. They were just the same helpful, cheerful, respectful bunch as always, and did their sniggering out of earshot.

As it turned out, some twenty miles away in Yeovil there was a scooter dealership, which claimed to have a replacement. Never having been to Yeovil, one of many towns with curious names, I welcomed the opportunity to penetrate its mysteries, and followed the A37 past Frome St Quintin and Ryme Intrinseca. All of Dorset is rotten with Roman history, and Yeovil is on an old Roman road, but by the time I got there the Romans had gone and they didn't seem to have left much behind.

There was still, I must admit, a high street, somewhat cobbled, and I had an hour or so to inspect it before the dealership told me that they didn't have the right key after all. Many of the shops were still functioning, despite the recession, and even Woolworths, empty and stripped to the bone, was still maintaining its ghostly presence. I also took in a very un-Roman bus station and a sixties shopping centre, but there didn't seem to be much else. I went back to Dorchester, wondering what marvels I had missed, but thanks to Wikipedia, I later discovered that Yeovil is chiefly famous for having had a really bad time with the Black Death and being bombed to smithereens in the Second World War.

But my time was not wasted. It was a useful outing and I was beginning to appreciate the qualities of my strange little steed. It's broader than a scooter and the feel is quite different. In fact, with my feet planted so much further apart, it was a little like riding a horse, which seemed pleasingly appropriate to my mission. It performs similarly to a bike, but because there are two wheels in front, rather than one, the lean is not the same. It's a kind of horsy lean, if you know what I mean. No? Oh, well. It was fun to ride, and I mastered the intricacies involved in locking and unlocking the two front wheels. As Wally said, they are most ingenious. When you are at a standstill you move a little lever and they lock in whatever position they're in. As soon as the vehicle starts moving they unlock automatically. As you corner they lean over in unison and there is no sense of having two wheels, but because they're smaller than a proper bike's wheels they bounce around more. They say it's almost impossible to lose them on a bad surface, and photographers covering the Tour de France are said to prefer them because they can get closer to the cyclists, but I never had that feeling of extra stability. For me that was the only other drawback to the MP3, but it was all in the mind, really, and easy to dismiss. All in all I was very happy with my choice.

Best of all, when I got back to the Beggars Knap I learned that the owner, a most dapper gentleman in the tradition of a gentleman's gentleman and with a butler's eye for detail, had spotted a coloured ribbon in the shrubbery alongside the garden path. And the ribbon was attached to my missing key. Feeling whole again and ready for my adventure, I pointed myself in the direction of Wem.

4

But Where's the Pie?

Could there really be a place called Wem? I know that even when I was eight years old and first heard the name I thought it odd, and that was a time when life was more than usually full of surprises. In 1939, soon after war broke out, my childish imagination was more urgently engaged by the prospect of German warriors parachuting down on to our roof-tops, swords flashing, to engage in knightly combat over the tiles and chimney pots of London. Big grey inflated elephants, called barrage balloons, were floating over the park, and when I took my new model sailing boat to the pond on Clapham Common, the local fire service turned out for practice and used it as a target for their hoses: my first sacrifice to the war effort. But all this excitement didn't last long. Before the end of the year my mother sent me to a school in Kent, and very soon after that it moved to a building near Wem.

There are remarkably few monosyllabic towns in England, and most have a strange cast to them – like Cark, or Claff, or Dull, or Dun, or Noke, or Nox, but for sheer intransigence I think Wem beats them all. The word suggests nothing and concedes nothing. Not Wemford, or Wemington, or Wembridge, or even Wemham. Just one singular secretive syllable, Wem, and I was drawn to its mystery. For two years, while London was being blitzed, battered and burned by Nazi bombers, I lived somewhere in the vicinity of Wem. And yet I had never been in the town and knew nothing about it.

It is on the map, of course, in a rather empty rural county called Shropshire, halfway up England on the west side. More word games, because there is no 'shrop' in my Oxford dictionary. Apparently, if you want to know where it all began you have to go back a thousand years to Scrobbesbyrigscir, and you're welcome to do so. But the map does reveal something about Wem. It is about ten miles north of the county seat, Shrewsbury, but the two big arterial roads that drive north from Shrewsbury seem to find Wem quite irrelevant and miss it on either side by miles. On the other hand, seven older minor roads radiate out from Wem, so it is obvious that Wem was once a much more significant place, a market town, perhaps, which has subsided in the modern era. Anyway, that was where I set my sights, to unravel something of the story of those two years, and to see if I could find traces of that older, emptier England of my youth.

I started out on Saturday, 10 July, and it just so happened that sixty-odd miles north of Dorchester was Bristol, where I had never been, and in Bristol lived a friend called Jacqui Furneaux who, a few years earlier, had put the normal process of ageing into reverse by getting on a Royal Enfield and riding it all over the world. She had offered to put me up and, even more importantly, to keep me company as we watched the final of the World Cup on Sunday.

So I pointed us north, back on the A37, carefully circling around Yeovil on the way. I still didn't have a decent strategy for avoiding the busiest roads. At first I'd thought that I should simply ask my TomTom to avoid motorways, until I found that there were countless other roads that used to be ordinary but were now really motorways by another name. Then I thought that if I deliberately rode out into the middle of one of those white areas on the map, full of spidery lanes without numbers, the TomTom would keep me off the big roads. Instead it seemed perversely determined to take me back to the nearest howling thoroughfare. Of course, I could have constantly fed new little village destinations into the machine but the mechanics of it were too cumbersome and defeated me.

Nevertheless I got a taste of what it might be like. I managed to ride through some delightful villages with names that would have looked even more enticing on a menu, like Champflower, Ditcheat, Chesterblade and Binegar. Then I rode way off the map and found myself in Vobster. (Who could resist the temptation to try vobsters with binegar?) I also discovered that in Vobster, you – but not I – can dive in a flooded quarry and inspect sunken limousines and underwater toilet seats. Then I realised that if I was going to get to Bristol at all that day I would have to leave my rural researches for another time.

I learned one thing from this trial run: I had been completely wrong about the English sense of curiosity. My plan to excite comment and conversation with my strange little machine was a complete failure. Even though I hadn't yet seen another bike like mine, scarcely anyone even noticed it. When I stopped at a pub in search of a pork pie, one person made an offhand remark, like 'Never seen one of them before,' but he certainly didn't invite confidences, and his tone suggested that he'd be quite happy if he never saw another.

There were no pork pies, either. Funnily enough, pork pies seem to have disappeared from British pubs. When I'm in America I never think of them, but in Britain I become mildly addicted. It seems extraordinary that, in a country said to be in the grip of an obesity epidemic, I can't find one of the pleasantest sources of superfluous fat. Pork pies and Scotch eggs used to be a staple of pub snackery. In fact, on my way to Bristol I wasted a good half-hour riding around Shepton Mallet looking for pork pies because my strung-out mind had got it confused with Melton Mowbray, where the best pork pies come from.

I spent a long time in Shepton Mallet because, by chance, I found myself behind a hearse, which was moving at a stately pace through an endless series of narrow streets, and I couldn't get past it without risking being in it.

Once I'd realised my error (Shepton Mallet would have been a good place to look for cider), I concentrated instead on the absence of interest in my MP3. I was really surprised, but by now even if I'd had the chance to change my mode of transport I wouldn't have taken it. How can I explain why this little contraption I was riding seemed so perfectly appropriate?

It has the innocence of a toy, nothing so purposeful as a motorcycle, let alone a car, yet at the same time it is a definite conveyance, a rolling seat, a vehicle out of a children's story – you get on it, and off it goes, like an ambulatory armchair. No gears to think about. All you have to do is steer it. I could have been Toad of Toad Hall, or Pooh Bear. I was wearing my elegant light tan leather jacket, my yellow kid gloves and a silk scarf. The only thing wrong with this picture is that I wasn't wearing my flat hat and goggles.

When I first heard of the MP3 it was believed by some that it could be classified as a tricycle, exempt from the helmet law, and I was overjoyed. But, alas, on examination, the two front wheels appeared to be not quite far enough apart. To the relief of those who love me, I wear the helmet. It definitely spoils the picture but, then, I don't have to look at it.

Jacqui's flat is right in the centre of Bristol, and looks down on the bit where they built over the waterway and now wish they hadn't. On the other side a little theatre was advertising, in huge letters, that *The Rocky Horror Show* was coming. Even after Jacqui had tried to explain, I still had no real idea what this is. Later she sent me pictures of the theatre crowd milling around on the pavement and I could see that it's a great opportunity for the girls to wear fishnet stockings and fake blood. What I missed entirely was that half the girls were actually men who enjoy wearing the same gear. I don't care to speculate on what that has to say about contemporary Britain, but what happens inside on the stage is still a mystery to me.

We went round the corner to an elegant bar to watch the World Cup Final, and it ended just the way I would have wanted. *Viva España!* We celebrated at one of the many stylish bistros in the refurbished wharves downstream, where I had a rocket salad with feta and olives, grilled sardines with toasted sesame-seed rolls, and a fine New Zealand sauvignon blanc. All around me, up and down the wharves, hundreds of others were enjoying variations on the same excellent, expensive fare. I worked hard at perfecting my Bristol accent, and delighted in calling everybody 'moy loverrrrr', and only afterwards did it occur to me that we were supposed to be in the worst economic depression since the early thirties. Back then only a pampered upper crust could have indulged themselves so freely behind plate-glass windows, while *hoi polloi*, with their pinched faces and shabby clothes, shuffled past enviously outside.

What's your point, Ted? Do you want everyone to starve just because the country's broke?

My mother told me about the hunger marchers who came down to London from all over England and Wales. I think it was from her I heard about a number of them walking into the Ritz on Piccadilly and ordering tea and a bun. It's hard to imagine today what a violent breach of decorum this represented. Men in boots and working suits, with braces and shiny bottoms, breaking into the sacred calm of this haven of aristocracy? Outrageous!

Apparently they were treated with respect, as one would a dangerous animal who might get upset and break the china. But the marchers were chosen for their vigour. The pinched faces were on the young men who attached themselves to the marchers as they came south.

I have yet more friends near Stroud so it made sense to wander up that way, staying well clear of the M5. It was easier to find smaller roads going north round Gloucester and Cheltenham, and I found myself suddenly among a flurry of Ingtons. I wonder what produces these name fads in such close proximity. There was Shurdington, and Uckington, and Tredington and Gotherington and Pamington and Fiddington, but what drew my attention initially was Boddington, because I like the beer. I wandered through it, hoping for a top-up, but, sadly, there was no sign of a brewery, just an ornate building with twisty Tudor chimneys behind an intimidating wrought-iron gate. A friend told me later that the beer comes from Manchester, and was famous in the nineties for causing models to speak with a 'Munchester' accent, but how was I to know?

The country opened up coming towards Tewkesbury as the swell of the land flattened out. It was a cloudy day, with a touch of drizzle, but I and the TomTom stayed dry. Somewhere near Witcombe I followed a sign to a Roman villa, up a long, winding asphalt track with a big, empty car park at the end of it. Then a longish walk got me to the villa, but it was something of an anti-climax: just a pattern of stones in the ground. Apparently in the third century this had been a great Roman house. The explanatory signs were good. An etymologist would probably have placed them in the sixties, because they described the house as having been connected to a 'Leisure Complex'.

As I went on north past Worcester, I began to notice how the countryside seemed to be divided into areas of greater and lesser prosperity, a kind of social zoning that was quite arbitrary, as far as I could see, since it was all very lovely. I often crossed the Elgar Trail, too, because the composer had apparently lived in a number of houses in the area. After Kidderminster I found myself on the A442, ten miles of sweeping curves with few interruptions, which seem to be a favourite with bikers: the road is peppered with signs saying 'Bikers Beware', and 'Ride Safe', and 'Think Bike'.

From time to time it occurred to me to experiment a little with the bike and see what it could do, and it was somewhere here that I found myself in a queue of traffic on a downhill stretch, having to edge along slowly. I thought, Since it will stand up on its three wheels when they're locked, why not just switch off the motor and let it roll down the hill under gravity? I found out why not. To my surprise, it rolled off sharply to the left and, before I could do anything about it, fell over. I was very impressed by the speed at which a van driver leaped from his seat to help me. My 'elderly gent' disguise was obviously working well. In fact I was fine – just on my side. There was a little mark on the paintwork, which no ordinary person would notice, but much later, when I revisited Stephen Burgess, he spotted it with satisfaction, as evidence that he had been right to buy something second-hand.

The day grew greyer and damper, and a pub called the Yorkshire Greys almost seduced me, but it wasn't even midday so I gave myself a stern talking-to, adjusted my attitude and persevered. Even on a cloudy, moist day the small roads were enjoyable, but I was having to deal with more and more traffic. I found myself lost again in my favourite meditation: trying to devise a car-free world.

Coming to the edge of Wem on a grey, drizzly day, it was not difficult to curb my enthusiasm. There was no reason really to suppose that Wem would be anything special. I passed between two strings of inoffensive two-storey houses, two Indian restaurants facing each other across the street, then an old church set back and a little above the road, its stonework tinged olive and green in the damp air, a variety of small shops, including one that called itself 'The Treacle Mine'. On the right the Old Post Office caught my eye because I didn't think anyone had built a new one in years. I found out later that it was now a pub. There were a few other pubs too, but most appeared to have closed down.

Then, half a mile on, the street bifurcated and my TomTom told me to take the leg that crossed a railway line at a level crossing. It was monitored by two ominous red cameras, twice the normal size, promising to take my picture as I smashed my way through the barriers to be crushed under a train. I don't know what the penalty for being crushed was. I do remember, though, that at the Regent's Park zoo in London the penalty for climbing over the fence into the lions' enclosure used to be five pounds. An enamel plaque on the railings said so. Of course, five pounds was worth a lot more in those days, and maybe lives were cheaper.

I managed to cross the rails safely and arrived at Aston Lodge where I had booked a room. It was late afternoon, and an attractively energetic woman of indeterminate age welcomed me with keys to the front door and to room number four upstairs. A large floppy dog inhabited the hallway and the lower steps. I walked over it several times during the following days, but without incident, and there were no cameras.

I splashed a little cold water on my face, tidied up, and came down to talk to the landlady, whose name was Denise, to see if she had any idea where Trench Hall might be, but she had only a vague recollection of having heard the name. It was too late to make any official enquiries. The town clerk had already told me in an email that a building with that name did exist, and had been acquired by the county council. I was sure that next day when I called on her I would learn more about it. So I set off into town, determined to view it in the rosiest possible light.

5

A School on the Fly

Back in 1939, my evacuation (and there seems to be no other word for it, unseemly as it sounds) was not typical. Indeed, if you were looking for a typical account of a little boy in flannel shorts, the kind that stick out a yard from his spindly legs, being sent off to God-knew-where with a gas mask, a suitcase and a name tag, then you should put this book away and read instead the work of my friend and classmate, Donald James Wheal. There was nothing at all normal about my case, and that was my mother's fault. The only thing she had in common with your smocked, mopped and slippered mum from the London terraces was that she was poor.

She was a German immigrant, who had arrived from Hamburg in the twenties, who fell for a charming Jew from Romania and, soon after I was born, fell out with him. Life at home must have been quite strained but I remember nothing of that. When I was five I was sent to Hamburg to be with my grandmother and aunts for three months, and it was only much, much later that I understood that I had been sent away during the complicated and rather sordid process that a divorce in the thirties entailed, my father having to be 'discovered' in bed with 'another'. For me it was a blessing, since I learned to speak German. Thrilled by the glamour of Nazi torchlit processions and splendid SS uniforms I was of course unaware that three years later, as the son of a Jew, they would have done away with me.

My mother became a divorced single mum in 1936 and, along with the rest of the British population (a sprinkling of aristocrats excepted), went to war with Hitler in 1939. This later involved bombing her mother and four sisters, all of whom happily survived, although the firestorm fused their cutlery into a solid mass.

By the time the war began she spoke faultless English, had British nationality, and had changed her name informally from Auguste to Anne to make things easier. Her life could never have been easy, but for a single mother in the thirties it was doubly hard. Her rebellious and determined nature embraced some strong political views, which must have brought her into contact with refugees from Nazi Germany. Most of them were Jews, but some were desperate political opponents of Hitler. I surmise now that this was how she found out about the school in Kent, where she sent me.

By the time I arrived there the school had about 150 children, and almost all had come from Germany, which meant that they were predominantly Jewish. Although some had been in England for several years, the majority had arrived quite recently and probably spoke no English, but I would not have noticed that particularly since I spoke German fluently.

We were in a fine mansion of Tudor origins called Bunce Court, but I remember nothing about it because, soon after I arrived there, the whole school of 150 kids and staff was wafted away from the south coast to Trench Hall, a large house in Shropshire. I wish I could remember more of those days. I have only the scrappiest recollections of small moments in a child's life. There was a smaller red-brick house about a hundred yards from Trench Hall where the youngest children lived, and for the first year I was one of them. A young blonde woman called Gwynne Badsworth was *in loco parentis* and I worshipped her. She showed me how to clean my fingernails by hooking my hands together in soapy water and massaging them, something I still do, which reminds me of her, and she taught me to knit squares for patchwork blankets, one purl, one plain.

I was daring. I climbed out of a window, worked my way along its ledge and climbed back in through another window. Or so I believe. I doubt that the window was more than four feet off the ground but I thought I was on a rock face. Between the two houses was a copse of beech trees, and I climbed to the top of one and looked out over the other trees around. About that my memory is very clear.

When I was moved later into the big house there were more jobs to do. An essential element of the educational methods employed at Trench Hall required children to spend a good deal of time doing practical jobs – *Praktische Arbeit* – which was just as well as there was nobody available to do the housework, and no money to pay them. Many of the jobs centred around the kitchen, which, as in most big houses of that earlier time, was in the basement, a mysterious place where the porridge was cooked overnight in hay boxes. One of the tasks we all took turns to do in the morning was to mix cocoa powder with milk into a smooth paste, and I remember how long it took and how hard it was to get rid of the lumps, something that puzzles me now because it seems so easily done. Perhaps we had to use powdered milk, which would explain it, but I remember being down in the basement kitchen for ages stirring that thick black lumpy mess.

We slept in bunks, one above another, and we had a game in which the boys on the top bunk would sit on the shoulders of the boys beneath and we would charge at each other like jousters. Being younger and smaller than many of the others, I was in the top berth and therefore one of the 'knights'. My partner in this jousting game was called Heinz, and it seemed to me that I had become so expert at mounting and dismounting my 'steed', with an almost magical fluency, that I was invincible. I have never been able quite to recapture that feeling, although the memory of it never left me. It is my version of the Golden Ball, which, according to the poet Robert Bly, 'represents that unity of personality we had as children – a kind of radiance, or wholeness, before we split into male and female, rich and poor, bad and good'.

If there was hardship I have no memory of it. We were fed and housed and kept as warm as was possible in those big, draughty old English houses. And, best of all, I believe we were loved. I don't remember ever leaving the house and its grounds, and the rest of the world was a million miles away. What I did not know – what I was now about to discover – was that I had been taking part in one of the most extraordinary rescue dramas of that whole hateful Nazi period.

Meanwhile, tonight, the delights of Wem awaited me.

6

Pub Fantasies

I suppose it says a lot about my limitations, but when I think of enjoying myself in an unfamiliar English town I think first of pubs. Surprisingly, perhaps, this has nothing to do with an unslakable thirst for alcohol. In America I never think of looking for a pub. In France I visit cafés, but not bars. It's just that I can't think where else, in an English town, I could hope to find a warm and welcoming atmosphere among good-humoured people. My imagination then rides easily along well-lubricated rails into a confection of gleaming brass and polished wood, a hearty blaze in an open grate, amiable patrons offering friendly greetings, the cheerful clatter of darts falling limply to the linoleum, steak and kidney pie or fish and chips, bar ladies with big voices and boobs to match, well, you get my drift. I can't stop myself conjuring up these seductive images, even though, to tell the truth to myself, I have had very little opportunity in the last decades to be part of them.

Before crossing the railway lines I had passed a pub, the Albion, that seemed to promise just such a heavenly scene, guaranteed to dispel the grey, moist weather. It had a well-kept façade, newly painted, with flower boxes at the windows, and a big red banner promising football to boot. I swung nonchalantly across the rails, cocking a snook at the cameras as I went, but, alas, the World Cup had ended the week before, and with it, apparently, the publican's dreams. It was closed and to let. The flowers were not for me.

I walked on to town, resisting the appeal of the 'John & Phoebe Ann Morgan Library and Reading Room', quite a mouthful of words to be engraved in stone over the darkened entrance. There were other pubs, flush with the pavement, not so nice, also shut down, but across the road was one that appeared to be in business. I won't say it beckoned to me, but it did let me in. It was called – well, to save embarrassment I'll call it the White Elephant. The door opened into a short passageway, with the toilets on the right. On the left another door let me into the saloon. It was rather small and shabby, with an L-shaped counter on my left, much more a counter than a bar. The two parts of the counter were about six feet long, and a post rose up to the ceiling where they met. It felt more like an office than a pub. My bit of counter was empty, but the other was crowded with three young men deep in conversation with the barmaid, who was about their age. They might have been at school together. I managed to drag her away from her admirers just long enough to sell me my pint before she went back to them. I would have liked to ask a few questions about Wem, but I was sure that if I ventured past the post I would be in hostile territory.

My impulse was to drink up fast and leave. I tried to find some interest in the notices pinned to the walls next to an old racing calendar, but they offered no nourishment. There was one other solitary drinker, a young man sitting at a table. He seemed oddly dressed – not that his clothes were odd but the way that he had buttoned himself up tight was strange. He had a canvas satchel clutched at his side with the strap still over his shoulder, and he stared vacantly over his pint at the wall opposite.

I'm at a loss to know why I have these exaggerated expectations of pubs. Where did I get this notion that there were always people in pubs ready, even eager, to make conversation? It certainly doesn't arise from my own experience. Most of my pub life was lived out in the saloons of Fleet Street, when I worked on the papers and where my company was ready made. I can't remember a time when I invited a stranger to join my smug little band. There obviously are people who know how to slither in, probably by proposing a large round of drinks, but successful bribery takes special skills.

A long-lost friend, Ronnie Collier, who was once features editor of the *Daily Mail*, told me about a trick he said newspaper bosses once used to boost their circulation in the thirties. They employed what they called 'sib spreaders'. For example, one of these men, working for the *Daily Mail*, would visit as many pubs as he could every night for a week or two (a career made in heaven), making multiple acquaintances and always carrying a copy of the *Daily Express*. Then for a few nights he'd go with his arm in bandages, complaining about a mysterious rash, and a few days later he'd appear miraculously cured, with no bandages, but carrying the *Mail*.

'It was the ink,' he'd cry. 'The ink on the *Daily Express.*'

Those fellows must have had a wonderful way with people, not to mention extremely absorbent livers, but perhaps pubs were different back in the days of the penny press and the penny pint. So far, on this journey, nobody in any pub had shown the slightest inclination to include me in their conversation, and I lacked the financial incentive to break in on their cosy chatter. As for ambushing those three tough young Wemmers, no way – so I left to pursue the essence of Wem elsewhere.

The Treacle Mine seemed promising and had a nice warm folksy feel about it. It was being operated by an obviously worthy woman who probably shared all my opinions and deserved my support. It was the kind of shop that sold almost everything I have never wanted (but perhaps should have). After inspecting everything and confirming that I didn't want any of it, I left with a slightly guilty conscience.

I could find nothing else of much interest in the street, but why should Wem be different from most of the other towns I had passed through? There was the Old Post Office, but I prefer my pubs to look like pubs so I passed it by and eventually arrived outside the two Indian restaurants facing each other across the high street. The one on my side, the Shabab Balti Centre, was clearly the classier of the two, and my class was feeling half empty. Besides, the other one, the King of Spice, had a disturbing sign in the window.

Hours of Opening
Sunday to Thursday 5 p.m.
Friday to Saturday 5 p.m.

Because I had recently made the terrible mistake of cleaning my glasses with one of those green scouring cloths, I peered at the sign more carefully, but I had not misread it. The information was painted, white on black. I can't say why this made me uneasy, but it did. I took a picture to prove that I wasn't hallucinating and turned into the Shabab, a stylish period establishment, rooted in ancient history by several massive and lustrous black oak posts and beams. A pleasant Indian in modern dress came to take my order and I complimented him on his magnificent woodwork. He agreed with me happily.

'Yes, they are very fine,' he said. 'They are plastic. See, you can knock on one,' and I heard the clack of an empty pipe.

The place was almost empty. Two women on my right were finishing their *biriyanis*, and at a table on the other side of the post two adolescent girls sat facing a boy, who was hidden from me but for his shoes and the back of his head. One of the girls, the prettier of the two, was in a permanent fit of uncontrollable hilarity, a spluttering, giggling, seething volcano spewing seamless mirth that surged to paroxysms of helpless laughter whenever a word was spoken by either of the others. It was plainly impossible for her to be eating even a mouthful. Occasionally, during a brief lull, she almost got the fork to her mouth, but a single spoken word would set her off again. I had the macabre thought that if the other two could keep it up long enough she might eventually starve to death, and die laughing.

I became desperate to see the boy's face. Why was he hosting these two girls? I guessed that the pretty one was his heartthrob and the other was the chaperone, but how was he dealing with this storm of caterwauling? Was he red-faced with embarrassment, or superciliously long-suffering, or darkly intent on prolonging the agony? The girl's performance was so phenomenal that I simply could not imagine how I would have dealt with it. Then, just as I was about to fake a visit to the loo so that I could catch a glimpse of him, it was over. They paid the bill, she fell quiet, and they left. He was just a bland, featureless seventeen-year-old with a funny haircut and no sign of PTSD. But you never know. Maybe that night he woke up screaming . . .

The streets were damp, dark and deserted as I walked back to the Lodge. Then I heard a distant, repeated howl, which resolved itself into a football supporter's chant. It came from the man with the satchel and the vacant expression, approaching me from a distance and wandering past me on the other side of the street. He was calling out, 'Come on, you Spu-urs.' The last drawn-out syllable dropped two notes in what I read as an acknowledgement of defeat. However many times he sent his mournful cry out into the night, neither he nor Tottenham Hotspur was going to come on. They were going down together.

7

You Can't Sell Fish in Nottingham

Next morning, in the room set aside for breakfast, there were three other men enjoying the 'full English' served up by Denise. One was her husband, who ate quickly and disappeared. Another was one of life's spectators, who munched and said very little, but the third was a tall, confident, middle-aged man with a driver's stoop called Nigel. His white van was parked outside on the gravel next to my MP3, and his business was delivering high-quality fresh food – mostly meat and fish, I gathered – to demanding customers. He was a fount of fascinating information.

'You can't sell fish in Nottingham,' he declared. 'Don't know why. It's all steak and chicken. Won't touch fish. Take it away! Sheffield's the same. No fish. But I can sell fish in Scarborough. Funny, isn't it?

'Prince Charles has eaten my prawns, at a house up here. He asked about them. "Prawns from Lancashire?"'

'You can taste 'em, see. They have taste.

'I sell to Wayne Rooney and his wife. And to Gary Neville up at Bolton. They've got a big spread up there.'

When I told Denise about my miserable pub experiences I unleashed a flood of reminiscence and dismay. Both she and Nigel had been publicans, and so, it seemed, had almost everyone they knew. The more they talked about it, the more I came to realise that owning a pub could well be the British equivalent of the American Dream. And I must admit I can feel the stirrings of that same ambition in my own soul.

Wouldn't it be great to inspire all that wonderful warmth and conviviality, to be the hub of the community, et cetera, et cetera? In my case it takes only a second or two to remember why it wouldn't. They are the same reasons why I was wise enough not to open a restaurant: I don't care enough about money to run a business, and I wouldn't much like half the people who came.

Several decades ago, in the fifties, there used to be a tiny French restaurant on Monmouth Street in London. There was tremendous competition for the few tables in this little shop, in part because the food was good value but mainly because, or so it was said, if the owner didn't like you he threw you out. That is the kind of pub I would like to run. The restaurant was called Mon Plaisir, and if you ate there it was at the owner's pleasure. Everybody wanted to be *persona grata* at Mon Plaisir, but I doubt that philosophy would work today.

Mon Plaisir changed hands a long time ago. Today it is bigger and much more elegant. There is little risk of being thrown out, and I imagine it is much more profitable, but if it were mine I'm sure I would ruin it, just as I would ruin any pub unlucky enough to have me in charge. Judging by the number of ruined pubs I had seen on my way to Wem, a good many capricious and unbusinesslike publicans must have fallen prey to the recession while pursuing the British Dream.

Denise and Nigel agreed that they would never go back into pubs. Denise said she was the one who had turned the Old Post Office into a pub but she had become disenchanted with the business. I think she said there once used to be thirty-nine pubs in Wem. Today there's a mere handful. But she did tell me there was a good one, the Raven, just outside town somewhere in a village called Tilley. And by chance, though I didn't know it, Tilley was where I would be going. But first I had to visit the town clerk.

The weather had brightened. There were patches of blue sky. Wem had a happier look. Through an opening off the high street I had missed the day before, I came into a small courtyard and saw the town clerk's notice on the door of a small but attractive brick building that must have been fairly new. I went into a light, airy space where a woman at a desk directed me up some stairs to an open gallery where Mrs Jane Drummond, the town clerk, worked.

I liked her immediately. How nice, I thought, looking back on all the dreary municipal offices I had once visited, that someone had managed to design such a pleasant working environment.

'This is a useful little fit-for-purpose office you have here,' I said, introducing myself.

She laughed. 'It used to be a café,' she said, smiling comfortably, 'but it works. Anyway, what can I do for you?'

'Actually I came to see you about the school at Trench Hall. You probably won't remember, but I emailed you a couple of years ago about Trench Hall, and you told me it still existed, that it belonged to the council, under another name. You see, I was there as a kid.'

'Oh, yes, I do vaguely remember . . . It's a boys' school.'

Her colleague called up from the ground floor with a question about a show, and Jane Drummond said, 'Yes, I'll take care of it, since I'm here,' and turning back to me, 'You know, you're lucky to find me. I shouldn't be here at all. We should be on our way to Iceland . . . We were on a cruise but it was cancelled at the last moment. But to go back to the school . . . I had no knowledge of its history until I went to a talk by a gentleman who lives at Tilley, who's done a lot of research into it. He told a fascinating story about the woman who brought the children across. I can't remember his name, but I'll just call someone in Tilley. She'll tell me. I know who you want to speak to. I'll have to do a bit of phoning around.' As she shuffled through her files, she added, inconsequentially, 'It's the sweet-pea show this weekend.'

'Is that important?' I asked.

'Oh, yeah!'

'Why?'

'Well, Henry Eckford, who created the modern sweet pea, had his nursery in the town. So we had a national show last year.'

I began to wonder whether my notion of Wem as a tiny place stuck in bad weather might need revising. Perhaps there was more to it than a bunch of decommissioned pubs.

'Well, it isn't very big,' she said. 'Five and a half thousand population, but we have a few claims to fame ...'

She agreed that it would be easy to think of Wem as a small place removed from the larger events of the world. In a way it was the story of Trench Hall that had opened her eyes.

'I was just fascinated by the fact that there was this little enclave of German-Jewish children. I'm not local, and you might think of this area as being insular – and parochial – when actually it wasn't. They had that school there, and GIs up the road, and all that sort of thing.' she said. 'But what was interesting about Trench Hall was that there didn't seem to be any mixing between the school and the local community. Do you remember anything?'

'No', I said. 'I had absolutely no contact with anybody outside the school.' Even as I was talking I realised I had always imagined that the school was buried far away from everything, in the depths of the country. But now, as it turned out, Trench Hall was probably quite close to the town. How had they managed to keep us so isolated?

Jane Drummond dialled a number.

'Oh, hi, Wyllis, it's Jane. You all right? ... What it is ... I've got a gentleman with me who's very interested in the history of Trench Hall School, and I remember going to a talk ... Yeah, but I can't remember his name ... Is it Reed, double E? ... Right. Thanks ... What? ... Yes, I'm still here. They cancelled it ... Oh, they tried to get us to go on another cruise, but we didn't want that. We were booked for Iceland, so we've taken a full refund and – don't laugh – we're going in a camper van to Ripon. But I was not pleased.'

She turned to me, laughing, and I tried for a joke about icebergs and Ripon, which failed miserably but she forgave me. Why Ripon? Well, there's Newby Hall, and the *Royal Scot* is parked there and runs every Sunday.

'Anyway,' she said, 'the name of the chap you need to speak to is Alastair Reid, at Brook Farm in Tilley. He flies helicopters at Shawbury, but Wyllis said she just saw him walk past her house so he'll probably be at home.' And she gave me his number.

8

'Collar the Lot'

Tilley is a small village on the edge of Wem, and the directions for getting there were peculiarly English, involving a church, a railway line, a bridge, a pub, and several opportunities for taking the wrong turn, in no distance at all. But I made the right choices and in fifteen minutes found Alastair Reid in the front room of his half-timbered cottage.

I saw a lean, fresh-faced man in jeans and a loose checked shirt. He had a shy smile, a friendly, unassuming manner, and snow white hair, which quite contradicted his remarkably youthful appearance. He was already at work in a purposeful way, laying out papers for me to look at.

Helicopters have not figured large in my life. Once, ages ago when I was riding through north-east India, I got to fly over the tribal jungle in a little Bell 47 – a glass bubble on a stick –

which was used to inspect oil pipelines, and I remember how thrilling it was to look down at the forest below. It used to be said that helicopters were difficult and dangerous to fly. I suppose by now they have become much safer and more routine, but I am still somewhat in awe of their pilots.

Recently a distant relation of mine by a previous marriage, who also flew helicopters as an instructor for the RAF, had been killed in one in England. He was a gentle and most civilised man, who took a delight in historical cartography and it was that, as much as anything, that brought him to mind. I imagine any number of military men settling gratefully into such peaceful pursuits, like Reid and his history project. My fellow died in an unusual and tragic accident: another helicopter had apparently 'sat down' on top of his. I'd heard about it as far away as California and mentioned it, thinking that Reid must have been aware of it, but I was taken aback when he said, 'John was my best friend.' It seemed extraordinary that we should stumble on this connection but I could have wished for a less tragic one.

When Reid had first set about investigating the origins of his cottage he hadn't known of the existence of Trench Hall, but when he found himself looking inevitably into the history of the village he could not but come across it. Trench Hall, it turned out, was right next to Tilley, even closer to Wem than I'd thought. It wasn't long before he had begun to ask about it, and although he described his discoveries in an unemotional way, they must have seemed exciting.

'I was very keen on history at school. Never on politics, though. They were always trying to teach us about Corn Laws, Reform Bills, and so on. I actually failed history O level.' But what he learned about Trench Hall was politics of a more dramatic sort.

'I found out about the *Kindertransport* – and got a couple of letters – and that sparked off this thing.'

Then he began to tell me about the series of extraordinary events, of which I was almost entirely ignorant, that had led up to my arrival in Wem. They centred around – and were often driven by – a remarkable woman called Anna Essinger. She had been born sixty years earlier in 1879 in a German town called Ulm. Anna, her five sisters and three brothers were the children of a Jewish insurance executive, but this was no ordinary business family. What apparently distinguished them was a lively humanitarian concern for society. They had reached out into the world, and when she was twenty, at the turn of the century, Anna had sailed to America to visit an aunt in Tennessee. From there she had found her way to Madison University in Wisconsin, where she had financed her degree doing teaching jobs and later managing a student hostel.

Meanwhile Germany was devoured by the First World War, and in 1919 Anna came home with a Quaker mission to provide clothing, hot meals and shelter to desperate people and abandoned children. Those were awful times in Germany. As a young girl my own mother lived through those terrible post-war years. She described the hardship often, and vividly – cold, hunger, misery, gleaning coal from the rail yards, trying to make meals of nigh inedible stuff. Those were times that drove people to extremes, either hardening their hearts to all but their own survival, or opening them to the suffering around them. The Essingers were clearly predisposed to help others.

Anna's sister Klara had already established a small school for handicapped children. Anna herself had been studying and practising the new educational methods of Maria Montessori, and in 1926, on 1 May, the family opened a boarding school in Herrlingen, just outside Ulm, with Anna as the head, and another sister, Paula, as the school nurse. To begin with there were eighteen children, but the school was successful and prospered. Anna proved herself an impressive headmistress and the numbers increased. But even though they insisted that the school should have no religious bias, the children were mainly Jewish, and as the country turned towards the Nazis it became obvious to them that their prospects were grim.

In 1933, the year that Hitler became chancellor, it was decreed that all schools must fly the Nazi flag on his birthday. Anna flew the swastika and took the children and staff on a day's outing, feeling that the flag flying over an empty school made an eloquent point. But she went far beyond empty gestures. Others might have drifted along, hoping for the best, but Anna was a fighter. That year she somehow acquired an old manor house in the south of England, called Bunce Court, and in August, as pressure on the Jews increased, she engaged with the parents in a dramatic scheme. Announcing that they were going on educational trips, she and her sister Paula went away with two separate parties of children, one to Holland and the other to Switzerland. And then they just went on travelling until they got to England.

One of her later English pupils, Harold Jackson, who became a journalist on the *Guardian*, wrote that the British, then wholly ignorant of what was going on in Germany, were astonished. Why had seventy-plus German children and their teachers 'suddenly erupted into rural Kent'?

Anna was, by all accounts, a typical German headmistress: short, stout, formidable, dressed in black, strict but kind, her eyes all but hidden behind thick pebble glasses because of the myopia that afflicted all her family. She was soon able to convince the British authorities that her school was an enhancement of the educational scene. She took on some English pupils, all teaching was in English, and the school connected well with the local society of Otterden.

All the time, of course, she must have been in contact with people in Germany, with Jewish relief organisations and the Quakers, as things gradually got worse until, in November 1938, Nazi hatred for the Jews burst out catastrophically in *Kristallnacht*. All across Germany Nazi storm-troopers smashed windows, destroyed synagogues and businesses, murdered many Jews on the spot and put another 26,000 into concentration camps where they almost all died.

Clearly all Jews in Germany and Austria were in deadly peril, and the news eventually filtered through to England. Within three weeks a plan by the Jewish Refugee Committee was negotiated with the Nazis and approved by the British government. Ten thousand Jewish children (and only the children) would be allowed out of Germany, each with one suitcase, a piece of hand luggage and ten German marks. It was called the *Kindertransport* ('*Kinder*' being the German word for 'children'). The scenes at those German railway stations must have been unimaginably painful, parents knowing they might not survive to see their children again.

A similar American plan to rescue twenty thousand children was scotched in Congress. Anti-Semitism was rampant in the USA, and the Bill was opposed on all manner of trumped-up arguments: one, straight from the Nazi playbook, claimed that Jews were just too ugly to be admitted *en masse*; another, that it was against the law of God to separate children from their parents. Better to let them die together.

Anna Essinger was naturally at the heart of the English rescue mission. She was at Harwich where the kids came over by ferry, and at Liverpool Street station where their fate was decided. She helped find homes for them and took as many as she could into her own school. It doubled in size, and the difficulties of absorbing so many traumatised children must have been considerable. Perhaps it was during those last months before the war that my mother came into contact with her. Being originally German and politically very active, she would naturally have wanted to help, although she was not Jewish. I have no personal recollection of any of this. Soon after war broke out she arranged for me to go to Bunce Court, and I realise now that it was something of a triumph to get me in.

The war brought a new set of problems for the school. Now every German in England was under suspicion. Churchill famously declared, 'Collar the lot,' and all the most recent arrivals at the school over the age of sixteen were whisked off to internment. Then the south of England was declared a Defence Area and the school was given three days' notice to move – extended to a week on appeal. An entire school, 165 people, in one week! Enter the Goodbeeyear brothers from Stockport. How they met Anna (and how they came by that strange name) I don't know, but they were Jewish and they happened to have a large empty house in Shropshire, Trench Hall, which they had bought at auction after the death of Colonel Nathaniel Ffarington Eckersley.

Boddington in Gloucestershire caught my fancy because I was hoping to grab a pint. The beer, alas, was away in Manchester, but I lingered to enjoy the Tudor chimneys.

Alastair Reid (below) outside his home in Tilley. It was researching the history of his cottage that led him to discover the story of the *Kindertransport* (see overleaf).

The cottage (left) where I spent my first year near Wem, and the big house, Trench Hall, where Anna Essinger finally found a refuge for her school of refugees from Germany.

Just in case you fancy a meal at the King of Spice in Wem, be sure to check the opening times (below).

1926 · BUNCE COURT SCHOOL · 1948

THIS
PROGRESSIVE
JEWISH BOARDING SCHOOL
WAS FOUNDED BY
ANNA ESSINGER M.A.
IN ULM, GERMANY, IN 1926
AND WAS BROUGHT
TO KENT, ENGLAND, IN 1933
AND EVACUATED HERE TO
TRENCH HALL
OVER THE WAR YEARS
1940–1946.

WITH GRATITUDE TO NEIGHBOURS AND THE PEOPLE OF WEM

KING OF SPICE

Hours of Opening
SUNDAY TO THURSDAY
5PM
FRIDAY TO SATURDAY
5PM

Just the sort of thing you would expect to see on the roads of Britain –
two hundred years ago, maybe.

Sloane School, Chelsea, the old school where I spent many interesting years dodging Hitler's flying bombs and various other missiles. But look what they're doing to my playground (below).

I persuaded an attractive young French girl on a bicycle to stop and snap me in Hardy Country (that's the Hardy monument at top right), and then went on to make a fool of myself.

The old market hall in Poundbury village (below) is actually new and causes traditionalists to gnash their teeth. But I liked it.

If we have to have roads, why can't they all be like this one, in Hampshire?

And here's the same thing again, in stone. Far from 'bringing me down', Winchester cathedral had me soaring.

A day at the seaside was my dream as a young boy. Here on the Riviera, at Torquay, you can choose between Agatha Christie and mackerel fishing. Who could ask for more?

This Eckersley belonged to an important family of mill and colliery owners in the Wigan area. He was a lieutenant colonel who had fought in the Boer War, and became a pioneer in the development of spinning machinery. He built three enormous mills in Wigan, the Western Mills, still visible today though partially derelict, but he settled eventually in Wem. He had four farms, helped to buy the recreation ground, funded scholarships, was a Justice of the Peace and, for a year, High Sheriff of Shropshire.

It did occur to me at one time to take my MP3 to Wigan because George Orwell's *The Road to Wigan Pier* was important to me in my youth. Probably it was my lingering disappointment in knowing that there could be no pier in Wigan that persuaded me against going. To me a pier meant a long boardwalk jutting out into the sea at a romantic holiday resort such as Blackpool or Brighton, crowded with amusement arcades, candy-floss stalls and factory girls on outings. I had never been to Wigan and I had assumed that the reference to a pier was a joke. I learned later that it referred, with some sarcasm I presume, to a loading dock on a canal that happened to be very close to the Western Mills.

I had read the book many years earlier and, aside from a powerful description of putrid tripe, I had only a general memory of the contents, but I knew they described the appalling living conditions of the miners and mill workers who laboured for people like Ffarington Eckersley – if they were lucky: in those days most men in the industrial north were unemployed.

It is the physical appearance of the book that has left a clear image in my mind. My mother must have subscribed to Victor

Gollancz's Left Book Club in the thirties because she had many books with that well-known soft red linen cover, and it is criminal of me to have lost them.

This redolent if rather fanciful connection between my mother and the upper-class f-f-f-f-f man fascinates me. I can summon up instantly the resigned bitterness on her face as she contemplates the hypocrisy of a man who sponsors recreation grounds and scholarships yet draws his wealth from families of eight crowded into two damp, crumbling rooms a hundred yards from an outside lavatory.

Of course, I can just as easily conjure up another Eckersley, who feels powerless to change the fortunes of thousands but is happy to find a small corner in Wem where he can contribute something useful. Indeed, it was a crippling housing shortage, born of rapid industrialisation, the First World War, and the Depression, that forced people to live in those dreadful conditions, and that might well have seemed to be an overpowering and insoluble problem. Yet other industrialists did face up to it. Like William Lever, for example, who built Port Sunlight, a completely new model village near Liverpool for 3,500 people; 800 houses with all kinds of facilities for the workers who made, among other things, the same Sunlight soapflakes we used in our kitchen before the miracle of detergents. But I have no doubt my mother would have held him in contempt, too, for his religious paternalism. There was no pleasing my mother.

In truth Eckersley did more. He gave a lot to Wigan as well, and at his funeral much fuss was made about his philanthropy. He sold his companies in 1918 and died, unmarried and childless, in 1935, just before Orwell wrote his masterful report on the condition of the unemployed working classes.

Of course the war, when it came, solved the unemployment problem quite handily. Later, when the men returned and, shockingly, threw out Churchill, it became obvious that things were going to be different, but I made my first forays into a wider England before much could happen to change the physical world. The landscape had scarcely altered since the twenties.

I was briefly infatuated (briefly, because she was several years too old for me) with Helen, a gorgeous blonde Scottish girl, and I hitch-hiked up the A6 to Glasgow to see her. She, her mother and grandmother lived in what today would be called a hovel on the Dumbarton road, beside the Clyde. The grandmother, as I recall, slept on top of the stove, but I thought little of it at the time. We were all used to living in tiny spaces. All my friends lived in rooms not much bigger than a dining-room table. My mother and I had the unusual luxury of four small rooms, including a bathroom, but only because she was willing to live on the top floor of a five-storey house, exposed to flying bombs. Nothing would persuade her to sleep in the shelter, and I won't say that the nightly pop-pop-pop of the buzz bombs lulled me to sleep.

Thinking about those days when I started to wander around Britain, I am struck by the cartoonish nature of that world. Because the classes were so distinct and remote from each other, it was simple to caricature them. A favourite cartoon character of that time was Colonel Blimp, who appeared regularly in the *Evening Standard*. He was a tubby little military dinosaur with a bald head and a flowing white moustache, who inhabited steam baths and was always wrapped in a bath towel. He uttered his inanities in David Low's cartoons, but he

didn't seem so far away from the real thing. The language difference between 'them up there, and us dahn 'ere' was far
more extreme than it is today. Even though I had schooled
myself to speak an acceptable middle-class English I occasionally met people from the upper crust whose vowels were so
contorted that I genuinely couldn't understand what they were
saying. It was still very much an 'upstairs-downstairs' world
and, near the top of my class in a grammar school, I was one
of the very few who could hope to climb the social ladder. I
remember vividly a physics master telling us, 'One day, some of
you may be thousand-a-year men,' as though he'd had a
glimpse of Paradise.

Trench Hall, the house that the Goodbeeyears offered to
Anna Essinger, was not very big, as country houses went, and
she had to find another home for a few pupils, but it seemed
stately enough to me. It had big french windows looking out
over a lawn, which sloped down to a ha-ha (a walled ditch,
invisible from the house, that kept the cattle away). There were
hedges along one side, and on the other a wood and thick
bushes of stinging nettles into which I was once unlucky
enough to fall.

One hot day in the summer of 1940, during what we called
the Phoney War, I was a little boy lying on my back on the
grass, listening to the hum of bees in the wild flowers and
watching a plane move lazily across the sky. Plainly imprinted
under its wings was the iron cross of the Luftwaffe. Why it felt
it could patrol the skies over England with impunity I'll never
know, but doubtless it was looking for targets to bomb.

'There were a dozen or so airfields within ten miles,' Reid
told me. 'Shawbury, of course, and Sleap. Then there was

Hadnall Station and Bridleway Gate.' He ticked them off on his fingers.

'People in Wem couldn't understand it. Why were these Germans free to wander around? A short bike ride could have got you to quite a lot of raw intelligence. There were prisoner-of-war camps too, one for Italians, one for Germans, but they were guarded, so why were these Germans free to roam around Shropshire?'

It was mystifying, and at first there was natural suspicion and hostility. It took a while to explain that there were Good Germans and Bad Germans, but gradually the atmosphere improved. It was the German staff, of course, who excited the most suspicion, but eventually the older boys, too, ventured out, helping a neighbouring farmer in exchange for milk. I never left the premises, except for one glorious fortnight when my mother allowed me to come back to London, during the Blitz, to witness London's docklands ablaze in the night sky. I'd bullied her into taking me to see 'Target for Tonight'.

Alastair Reid's directions were simple to follow and I found myself looking at the house seventy years later. It was now a school for boys, and a conference centre had been added. Broad sweeps of asphalt surrounded it, the ha-ha appeared to have been filled in and, not surprisingly, I sat on my Piaggio in the driveway feeling rather alienated from it. My memories, after all, had little to do with the building itself. It would have been through that french window that they carried me from the stinging-nettle patch to lay me out on a table and slather me with calamine. And behind that door would have been the hot-water radiators we warmed our frozen hands on after playing in the snow – to our later regret when agonising chilblains formed.

There seemed no point in asking to go inside. Everything would have been changed and painted over. The brick cottage where Gwynne Badsworth had watched over me was still there, and there were a few trees left between. I photographed the blue plaque on the wall: 'This progressive Jewish Boarding School was founded by Anna Essinger MA in Ulm, Germany, in 1926, and was brought to Kent, England, in 1933, and evacuated here to Trench Hall over the war years 1940–46. With gratitude to neighbours and the people of Wem.'

How little this says of an extraordinary life of diligence, courage and duty to humanity. Of course, I entered her life only briefly and by chance. Alastair Reid showed me a list of men – artists, academics and writers among them – who might never have survived their childhood without her foresight and determination, but of all the old boys of this school, the one I am most delighted to have been associated with was also one of the funniest men of his time: Gerard Hoffnung. And whoever has not heard him deliver the Bricklayer's Story to the Oxford Union should rush to Google right now.

Accidentally Reid also opened another window on to these isles that might otherwise had stayed shut, when he talked of a dozen or so airfields within ten miles of Wem. At first this seemed scarcely credible, but when I found the Bones Aviation Site on the Internet I learned that there were almost 1500 airfields in use in Britain during the war, which translates roughly into one airfield for every sixty square miles.

9

Dodging the Rain

Instead of my life becoming simpler as I grow older, I find myself entangled more and more with a mindless corporate bureaucracy. To some degree, of course, it's my own fault. I succumb to offers and fall victim to the fine print, but more and more often now it is the inexplicable incompetence of huge corporations that wastes my time, costs me money and leaves me fuming with no chance of redress. In those rarified strata where they plan their mergers and acquisitions, where they chart their strategies, concoct their evasions and launch their products, they are highly efficient, no doubt. The devil is in the interface. When it comes to dealing with the public they are absurd, probably because they think they're faultless and any customer with a problem must be an idiot. What better way to defeat an idiot than with another idiot?

But on the day I left Wem it was not Business, Big or little, that was consuming my last reserves of patience: it was everybody's natural enemy, the tax man – or rather the tax woman. She had a quantity of my money and, by refusing to answer telephones or emails or letters, seemed determined to hold on to it. Her fortress, according to the letterhead, was Ferrers House in Nottingham. Normally I sit impotently outraged in the wilds of northern California, with a one-hour window between the time I get up and the time she leaves her office, but now I was less than two hours away from her stronghold and, because her phone remained permanently busy (With what? I asked myself. Certainly not with humble petitioners like me), she had no idea of the monster slouching towards her Bethlehem.

Quite where this fits into my story I'm not sure, but it seems to me that a comparative study of government bureaucrats, then and now, must have its place. Back in my youth, they lived and laboured in brown warrens that had been burrowed out of rows of large, anonymous buildings somewhere within a pebble's throw of the Thames. Behind brown doors, under nicotine-stained ceilings, brown files accumulated on brown wooden desks bearing brown stains from brown tea, delivered often enough by brown tea-ladies from the decaying Empire. Clerical life under King George VI was brown. But now they had been seized, as if by rapture, and transported to Nottingham. I was in some excitement about what I would find there.

Because Wem was trapped in a nineteenth-century time capsule, as far as roads were concerned, the route across to Nottingham started on those same lovely quiet country roads that I wanted my TomTom to discover for me. It was a blustery day, with the wind blowing rainclouds away before they could catch me. I rode from village to village in the general direction of Uttoxeter, between big bushy hedges and slightly unkempt fields, spotting occasional large houses set back behind trees. One was a baby castle, somebody's folly, named the Citadel, but for the most part they looked serious and important, dark and damp. For an hour or so I enjoyed the sensation of having escaped to an earlier age before I was forced back, cursing, on to the big highways.

It was about midday when I started winding through the even bigger motorways around Derby, and just before I got to Nottingham that the rainclouds zeroed in and hit me with a shower. I arrived at Castle Meadow Road in a sudden burst of sunshine, but wet under my rain clothes. I had no particular expectations, but Ferrers House astonished me all the same. The word 'house' is quite misleading. I found myself about to enter a bureaucrat's paradise, a small city of elegant multi-storey buildings protected like any imperium by a guard house, a boom, and an officer in some sort of pseudo-police type uniform. He allowed me graciously to park my MP3 on the pavement so I took off my waterproofs, then walked into this Palladian celebration of taxation. He directed me to a reception office, but on my way there I saw to my left that one of the buildings invited 'Enquiries' so naturally I entered it to enquire.

On the left of the entrance was a small group of huge women – I mean truly mountainous – smoking cigarettes. The thought struck me immediately, before I had time to censor it, that they had no right to be mountainous in such idyllic surroundings, and I wish I didn't feel compelled to reveal the unpleasant truth about my prejudices, of which I have many more. Well, everyone has them; the important thing is to be aware of them, know them for what they are and limit the damage. I wish I could say that some of my best female friends were mountainous. I can't think of a single one, but I certainly wish them no harm, wherever they may be.

The entrance was a huge steel and glass tent, like the atrium of some Arabian palace hotel, and facing me across a broad expanse of fancy flooring was a long bench where several more pseudo-policemen, all with angular jaws, sat in a row, like a panel of judges as seen by Alice through the looking glass. Strangely, booming out from behind them and filling the whole building with a loud, echoing racket, there came the sound of balls being whacked and cries of triumph or despair. It occurred to me that maybe this was a rehab facility for tax people overburdened with guilt – a kind of detax centre.

I applied to one of the judges for a licence to speak with my tax lady, and they went into conference for a while, consulting various files, before telling me that she wasn't there any more, and in any case she was at lunch, so I said I'd wait and watch the people come and go.

I was impressed by the number of men who came through the revolving doors, then stopped and looked around with an angry expression to see who might be belittling them, obviously a stage in the detaxing therapy. After a while I was called back to the judges' bench and handed a telephone receiver. A woman who was not my tax lady said that she had moved on and they were re-organising, and that they had indeed received my letter of two months previously but couldn't find it because in the process of re-organising they had thrown all the papers up in the air to see where they might fall in the hope that this would suggest a better form of organisation than the one that had failed them so far. Well, to tell the truth, that was my fantasy, but it seemed as good as anything they were working with. However, I will say she was very nice and sympathetic, and I could tell that the therapy was working. I couldn't get to see her, though, and I'm glad, really: she might have been mountainous.

To be perfectly sober for a moment, I have to admit that I was shocked by the opulence of this palace of revenue. Here we were in the middle of a terrible financial crisis caused by overheated dreams of perpetual profit, and if any corner of the economy could be expected to hold out against wretched excess you might think it would be Her Majesty's Revenue and Customs. But the worst of it, from my point of view, was that for all the squash courts and fine clerking they still couldn't find my money.

When Orwell was writing in the thirties he pointed out that whatever the socialists said about the evils of empire nobody really wanted to get rid of it, having grown fat on it all those years. The alternative, he wrote, would be to 'reduce England to a cold and unimportant little island where we should all have to work very hard and live mainly on herrings and potatoes'.

Well, we had got rid of the empire and so far, somehow, we had staved off the ugly fate he had foreseen without working very hard at all, but I had a feeling it was coming. Fortunately for me, I'm rather fond of herring and potatoes. Unfortunately (as I was to learn further down the road) the herring was already on its way out.

*

My idea, having bearded the lady in her den, was to go south again in the general direction of Bedford and on the way I thought I'd drop in on Meriden, and revive some more old memories.

Still hoping to find a route away from the main roads I wasted a lot of time dodging off into inviting little lanes only to find myself forced implacably back on to the highway. The weather – my enemy – looked awkward, and I could see I was fairly sure, sooner or later, to get wet again. Of course, getting wet is a big part of biking, but I couldn't bring myself to think like a biker. I clung to the illusion that I was a gentleman of leisure, ambling across the countryside, and such a person travels in sunshine. He would not be seen cocooned in waterproofs. The rainclouds were interspersed with patches of blue and I rode on determined to slip between them. And, in fact, I managed quite well for about twenty minutes, until a great ugly black mass reared up in front of me. With my tail between my legs I scooted back to the last petrol station and waited until the doom-laden cloud had passed over.

When I'd first conceived the notion of riding a motorcycle around the world, in 1973, I had never heard of Meriden. In fact, when I think about it, it is remarkable that I could have been a successful editor on a national daily newspaper and known so little about my own country. I suppose that when I complain about the state of journalism today I should bear that in mind. There was a time earlier in my career when the *Daily Express* wanted to ship me off to Manchester to have my edges buffed down on the *Manchester Evening News*, but I refused to go. I have to admit now that it would have been good for me, but

I had listened to bloodcurdling stories about its fearsome news editor, and had a craven suspicion that I might not survive the experience.

Not that knowledge of Meriden itself is an essential qualification for citizenship. It's a village of a few thousand souls squeezed in alongside Birmingham and Solihull, and until the Second World War its only claim to fame was an old monument asserting that it was the geographical heart of England; a claim that later turned out to be false anyway. But when Triumph Motorcycles were bombed out of Coventry by the Luftwaffe, they set up shop at a factory in Meriden. In the post-war years British motorcycles still ruled the world, but after a couple of decades complacency eviscerated them, and it was in their dying days that I collected my bike.

I won't forget the excitement and trepidation I felt at that thrilling time. Triumph put me up in a pub down the road. The Vietnam War was in full swing and the soundtrack accompanying my Meriden experience was an endless loop of 'Tie A Yellow Ribbon (Round The Ole Oak Tree)', which I mistakenly assumed was about soldiers coming home. So I thought I'd see what it felt like now, and brave the curses that might be hurled down on my funny machine from the celestial keepers of the Triumph soul.

But instead of curses they rained down drizzle and confusion. Nothing was as it should have been. The factory would have been my point of reference for finding the pub, but I was soggily unaware that the factory had been demolished and turned into a housing estate. I searched fruitlessly for anything to jolt my memory. After rejecting two rather high-priced hotels that offered nothing remotely interesting, I dug out my

B&B book, discovered what seemed like a good deal and wasted a lot of time riding in the wrong direction.

I found it at last and counted myself lucky that it wasn't a bug-infested whorehouse, because I would have accepted anything. I was greeted by a man who was almost certainly older than he looked. He was surrounded by a party of visiting teenage boys, who he fussed over like a scoutmaster.

When they'd gone wherever they were going he spoke of their virtues with shining eyes. 'Marvellous kids,' and 'They don't make 'em like that, these days. Aren't they great? Lovely kids. Just what this country needs,' and then, almost without missing a beat, he told me, 'I don't mind admitting that I voted BNP this time.'

'Well, that didn't do much good,' I said, as cheerfully as I could. He was among a half-million or so who had voted for the British National Party in that year's elections. They had failed again to put anyone in Parliament, but it was their best showing by far in twenty-seven years. A bit disturbing, I had to admit. He was not to know, of course, that embedded in my DNA is a profound dislike of anything even remotely tainted by Fascism.

Still, he was the first specimen of his ilk that I had encountered. I was curious to know how anyone could be attracted to a party born of Oswald Mosley's British Union of Fascists and his sinister attempts to ape Mussolini and Hitler. But I didn't have to ask him. He was anxious to tell me, and his beliefs flew at me like a flock of tweets.

'This isn't Britain any more,' and 'The British working man has no chance against all these immigrants,' and 'There's no opportunity left in this country,' and even, 'I've nothing against them personally – they just shouldn't be here,' all uttered in a spirit of sweet reasonableness, though there wasn't a lot in the way of reason. Pretty much the only thing he didn't say was that some of his best friends were Jews. And yet I couldn't dislike him. He was obviously an energetic, frustrated man up a blind alley, probably having a hard time keeping things going. We talked about the state of the nation for a while, and I was quite diplomatic for fear of being thrown out.

He told me about a good pub, and I wandered off down a country lane under a light grey sky between moist, grassy verges and companionable trees, all properly restrained and comfortably proportioned and very English, with not an immigrant in sight. I got lost alongside a small lake where a man in a mildewed cape, dismantling his fishing rod, set me on the right path to the Bull's Head. It was a very good pub indeed but, like most good pubs these days, it was more of a restaurant, with couples sitting at separate tables minding their own business: much decorum but not much conviviality.

In my cosy little bedroom that night, I thought about where to go next. My course through the country had been fairly vague so far but there was a rough sense of chronology behind it. I liked the idea of starting with Wem and moving on through the years, and in fact the next important stop in my childish appreciation of the English countryside would take me back to the West Country, but since Meriden was on the way I had hoped to profit from it, and after Meriden there was Bedford, or rather the nearby village of Cardington, where famous British airships like the R101 had once been hangared.

10

Boots, Ladders and the Bomb

RAF Cardington was where I arrived one day in the winter of 1954 to begin my National Service. Everything was quite civil. I got my uniform and my boots – a big deal, the boots – and various bits of equipment. Then I was transported north to a place called Padgate. None of the rumours had prepared me for it. I stumbled off a bus, propelled by the bloodcurdling screams of a band of raging psychopaths. They later turned out to be quite reasonable young corporals in the Royal Air Force, but in my eyes at that moment they were ageless agents of doom. Never having been screamed at in my life before, I endowed them with the power of life and death. I was literally traumatised, and from that episode I learned more about life and politics than from all my years of reading. I was twenty-three years old, and when I discovered a few days later that they were only eighteen, it came as a terrible shock to realise how easy it is, given the right circumstances, to overcome reason with terror.

I did my square-bashing in the frozen north of England and, once I had recovered my equilibrium, I found, to my amazement, that it was quite exhilarating. I would never have thought that marching around in the snow and swinging my arms would appeal to me, but I had the good sense to realise that it was actually doing me good. By the time it was over, after two months, I was probably fitter than I'd ever been, but that just made what followed seem even more of a waste. I was sent straight back to Cardington to hand out test sheets, and endure a few months of mind-numbing boredom and quantities of NAAFI tea, spiced, it was rumoured, with sex-drive-inhibiting bromide.

I began my air-force career as a recruit, and continued later as a rather grandly and ironically named Personnel Selection Assessor: grand because I had no rank, and ironic because it was the last thing I ought to have been doing. My job, as a Leading Aircraftsman Plonk, was to hand out test papers and tell people when to start and stop writing. Thus I became a lowly cog in the machinery designed to ensure that everyone in the air force was given maximum opportunity for personal growth by assigning them to the work for which they were least suited so that they would learn to make the best of a bad job. What it did do, in fact, was to encourage a sturdy resentment of all forms of authority and a determination to do no job at all, if it could be avoided.

The preferred method of avoidance was to walk around in a resolute manner with an important-looking piece of paper called a chitty. Two years of doing this produced what you might call the Skiving Generation. The up-side was that some people went on later into music, peculiar fashions and drugs and became the Swinging Sixties. The down-side was that people armed with clipboards and chitties found it easier and more interesting to organise others, tending to become shop stewards and, according to Mrs Thatcher, to bring a once-proud nation to its knees.

But I was luckier than most. Eventually I found a third way, grasped the keys to the kingdom and bent the RAF to my own devilish schemes. Well, it wasn't quite so Machiavellian, really. I persuaded a well-meaning squadron leader that the recruits would be happier if they had a magazine to entertain them, and that I was the one to produce it. So *Scramble* was born. Happily this required me to be in London most of the time where I discovered sympathetic ex-RAF types, like Peter Sellers, who enlisted his mates from *The Goon Show* to help me fill the pages.

I also touted for advertising, and my most memorable deal was with a firm of military outfitters. Among their uniforms and accessories they had a product consisting of a flat bottle with a tube that could be strapped to the thighs of mature officers on parade. There was a picture of it in the magazine and it was called the Y-B-Wet.

I did, of course, have to be on the station some of the time, but my life there was immeasurably easier than most. I was allowed to have a separate little room for my bunk, usually allotted only to those fearsome corporals. I was excused parades, so I could wallow happily in the rubbery sheets with which we were issued while others were stamping their feet in the frosty morning. I was given a little office space as well, and enjoyed the work immensely.

Getting to London wasn't so easy. Taking the train was a long-winded affair. There were very few cars on the road in 1955 – maybe one for every five families – and a lot of those were in London itself, but we had a priceless asset in our hut in the shape of a burly baronet. Generally speaking, baronets became officers but Sir Harry, by his own admission, was so deficient in officer material that he remained an LAC Plonk, like me. Unlike me Sir Harry had a limousine, which he drove, when they let him out, to his home in Golders Green, and occasionally I got a lift.

At the time, of course, as we sailed down the A6 in the Austin Baronmobile, everything appeared normal, but in my mind's eye I can still see the open road. Imagine it. An open road. In England. Not just a gap in the traffic. Emptiness. Ah, here comes a bus, all alone. It stops. It starts again, and disappears in the distance. More emptiness.

On this journey, as I ride around on my little praying mantis, I curse the traffic. There are times when I see the endless onrushing horde of cars hurtling towards me as a Hitchcockian nightmare in which blind metallic things have taken over the world. And they have, and it is a nightmare. Even though the people in the cars think they know what they're doing, where they're going and why, the sum effect is mindlessness. Outside the metallic skin of the car, at a different level of perception, it is all insanity. No intelligent being would ever design a social order in which hundreds of millions of people are forever dashing around in boxes on wheels.

And me? Well, I'm outside the box, and I'm not dashing, but still I am part of the problem. It seems to me that I've been obsessed with the problem half my life. The trouble is, I can imagine a completely different way of doing things, several different ways, but as the Irishman is supposed to have said, 'You can't get there from here.'

From my point of view, *Scramble* was an outstanding success since it liberated me from all that military nonsense. I was kept pleasantly occupied for eighteen months, I learned a great deal and I was very diligent in my duties. Of course the magazine lost money, but not too much, and it was paid for out of a kind of airmen's welfare fund whose name I have forgotten, though I think it included the words 'President' and 'Institute'. When my time in the RAF was up, I tried hard to get a commercial publisher to take it on, and for a while I thought the famous Reverend Marcus Morris, he of the *Eagle* comic and the creaking corsets, might do it, but he failed to grasp its dazzling potential, and it died after me.

Then, just a few weeks before my demobilisation, I was stuffed back into my uniform and made to go north again, like a proper military person, to learn to be a fire-fighter. Once again I was seduced by the manly physical life. I learned to run up and down ladders with people slung over my shoulder, and I was almost paralysed with mirth when a rather inept colleague, trying and failing to join up two hoses in time, was drenched with a mixture of water and slaughterhouse juice – that being a foaming agent in use at the time.

But the whole exercise had a darker side. We were being trained, in fact, to anticipate an attack by atomic weapons. We were given lectures showing how much of London would be vaporised by a well-placed bomb and told we would for ever after be members of the H-reserve. It was then that I was given an instruction so idiotic that it has made me treat all official instructions since as the babble of idiots. Apparently the authorities believed they would get twenty-four hours' notice of a likely strike, and we – the noble H-column – would be told to leave London immediately, and surreptitiously, and to assemble at Moreton-in-Marsh. *In no circumstances* were we to tell anyone why we were leaving.

Oh, right! Like, 'I'm just popping out to the movies, Mum. Don't stay up. I might be home late.'

Then, when the dust had cleared, we would return to put out the fires and clean up the incinerated remains of our nearest and dearest.

Around that time my stepfather provided me with another priceless example of the same idiocy, from the Underground. He said that waterproof doors had been built into the tunnels running below the Thames so that if the Bomb breached the tunnel under the river the entire system wouldn't be flooded by radioactive river water. Unfortunately, though, they had to be manually operated from the wrong side of the door.

'After you, Claude.'

'No, after you, Cecil.'

Scramble got me a job in the papers. Through a girlfriend I had come to know Andrew, who worked in the art department of an advertising agency, and he had offered to design the covers for me. He also happened to be the son of the editor of the *Daily Express*. This editor, Arthur Christiansen, was a redoubtable figure in Fleet Street, and the *Daily Express* in those days was a phenomenally successful and powerful newspaper, selling around four million copies a day, all without the help of naked girls on page three. So, Christiansen saw my magazine, I got my break, and I fulfilled the dream of my physics master and became a thousand-a-year man, but by that time twenty pounds a week only just got me through the door.

Still, it was enough to get me into motoring madness. I bought an old car, and this was not just any old car. This was a 1937 Lancia Aprilia, a famous car in its day with many remarkable innovations, which I learned about as I worked on it outside our house in Kensington Park Road. Sometimes I was out there all day, and it was the sort of street where distinguished older men strolled by a lot and beamed at me.

'I say, you're a lucky chap. What a fine old thing she is. Why, I remember . . . '

And indeed she was a fine old thing: the first with an aerodynamic design by Farina; doors that opened without a central pillar to reveal delicious cream leather upholstery; aluminium lining on a monocoque body, and so on. Unhappily, she didn't go – well, not much, and not far. Eventually, as I slaved under the bonnet, I got tired of my role as a roadside catalyst for old men's reveries, and by the time I found out what the main problem was (hairline cracks in Bakelite plug extensions that couldn't be replaced) I'd had enough.

I bought an Austin-Healey Sprite, new, for £500, and drove it all over Europe, but by then the traffic was thickening and in the early sixties it became lethal. When I drove out to Essex at the weekends to visit my mother, there was always a place, somewhere along the A12, where we slowed to a suitably funereal pace and passed, in single file, the bodies laid out on the median strip, waiting for an ambulance. That was when the first lurid 'Keep Death Off the Roads' posters appeared, with graphic photographs of gruesome accidents and the miseries of bereavement.

Those were also the days of the lamentable Beeching Report, sponsored by that cheeky chappie Ernest Marples, Conservative minister of transport, who probably had his tongue in his cheek as he shut down the railways and promoted the motorways: it was his company, Marples Ridgeway, that built the M1, the Hammersmith flyover and various other monuments to motoring. When it was pointed out that there might be some conflict of interest here, he sold his controlling shares – to his wife. The scandals that embellished the Macmillan government of the sixties usually bring to mind poor John Profumo and Christine Keeler, but Marples exceeded them by far in his depredations. Eventually tax fraud forced him into exile.

It's fun to resuscitate all these juicy shenanigans, but beyond the laughs, there was serious damage done as we diverted more and more of our national life on to the roads and motorways of England at the expense of a much more conservative and eco-friendly system of travel. In this respect we have followed the USA more closely than others in Europe, and, with a far denser population, the stifling effect of congestion in the UK is much more marked.

Over the next ten years, as the number of vehicles on the roads trebled, the government bore down heavily, trying to stop the slaughter, but there was great resistance. It was another twenty years before seatbelts were made compulsory and the British driver was finally tamed and trained.

Today the volume of traffic on British roads is ten times what it was when I started travelling around, and the consequence for drivers in this small, densely populated island was inevitable. Most driving in Britain today is like navigating a board game – snakes and ladders but without the ladders. On the main road system, at junctions and elsewhere, the surface is a grid laid out with multi-coloured lines and symbols, and every square confronts the driver with signs imposing new rules and threatening new punishments. It has happened gradually, and I suppose most British drivers scarcely notice it, but coming from the wide open freeways of America it hits me hard. Today the control exercised over the average journey is so thorough that it is now quite reasonable to expect thinking machines to do the driving for us, and there are plans to make that a reality.

And in truth, if I were an average British driver, I would find it a huge relief. There's not much pleasure to be had from driving a car in these circumstances, unless you enjoy playing dangerous games and risking huge penalties. I'd love to have Big Brother as my chauffeur (George Orwell is always creeping in). Just think how pleasant it would be to stagger out of the pub into your automated capsule, push a button and fall into a snooze until it deposits you at your doorstep. Well, I'm sure they'll find a way to spoil that dream.

I wonder, though, what the cumulative effect of these gradual but ever tighter controls over the years – and not just on the roads but in general – have had on the British psyche. It seems to me that if you deprive humans of their accustomed freedoms they will find others and, with the mounting affluence of the past decades, there have been great new liberations: the freedom to travel and the freedom to shop are two that spring to mind. During my last years in England I enjoyed both to the full, so I realise now, and not for the first time, that I lived the prime of my life in a golden age, which is gradually fading to grey.

Travel is becoming ever more entrammelled by a frustrating bureaucracy. Shopping has certainly taken a big hit. If the desire for freedom is incompressible, where will it bulge out next? Alcohol, drugs, food and sex are good, but they hardly count as self-expression. I can see why my last night's host might be looking desperately for someone, like the BNP, to give him back the freedom he thinks was stolen from him, even if he can only have it at the expense of someone else's. He probably doesn't have the imagination to see that his would be next on the chopping block. That's the way of the Fascists.

When I looked at the map I saw that if I took a circuitous route to Bedford I could pass through Moreton-in-Marsh. I'd never been there because, thank heavens, I had never been called upon to perform my grim official duties, but I was curious about the place and I saw that if I took this roundabout way through the edge of the Cotswolds I'd have a better chance of finding myself on smaller roads. The weather was fine again. I aimed my TomTom first at a tiny village called Aston Cantlow, which got me across the M40 and other major motorways, and then I pointed it at another, Cow Honeybourne, working my way south to the west of the River Stour through ever more lovely hamlets until I came down through Chipping Campden and past Batsford to Moreton-in-Marsh.

Like everyone else who has stories to tell, I have told the story of H-column and my life as a fire-fighter many times in the past, so Moreton-in-Marsh has often been on my lips, but it never occurred to me to wonder what the place was really like. To me it was as fictitious as the imaginary RAF station Much-Binding-in-the-Marsh, which it inspired, and which was the locus of a radio comedy that had the nation in fits of mirth from 1944 into the fifties. So it came as a shock to see what a beautiful town this was. I found myself on a broad street, probably designed to accommodate passing flocks of sheep, but the buildings that framed it also civilised it. From three or four storeys up a whole sequence of honey-coloured stone façades, beautifully fashioned, gazed down benevolently on sophisticated shops and restaurants. But beyond beauty, it was also a town I felt a strong and immediate sympathy with. I have always had an affinity with stone, the warmth and golden glow it sheds in a hot sun, and I wished I could have stayed awhile, but I had a room booked in Bedford, and I was discovering that rooms were not so easy to find.

The RAF station of that older, darker time was now a huge establishment entirely devoted to fire-fighting. No doubt Kenneth Horne of Much-Binding would have found plenty to laugh at, but I kept my pyromaniac tendencies at bay and passed it by. What I didn't know was that Moreton-in-Marsh was in the squiredom of Lord Redesdale, who lived at Batsford House, where he brought up his six batty but brilliant daughters, the Mitford girls, one of whom married Sir Oswald Mosley and worshipped Hitler, just like my aunt. Funny how these things keep coming round.

11

The Biggest Indian in Europe

The only bed I had been able to find in Bedford was at a B&B called the Elms on De Parys Avenue but that suited me quite well. Back in my RAF days, when I was still governed by a rigorous programme of parades and inspections and other martial conceits, I spent the few free evening hours I could get in Bedford, at the cinema or visiting a friend. Cardington was only few miles to the south and there was a bus service. Often the timing to catch the bus was tight, and to get back to the billet from the mess hall and then to the bus stop in time was a rush. We were all issued with eating utensils, collectively called 'irons', and one evening on the bus, noticing that others were glancing at me rather oddly, I looked down to see that I was still grasping my knife, fork, spoon and mug in my clenched fist. Those were the days when I could still hold on to things. Nowadays I have no idea where I've put them.

The friend I usually visited was Stanley, a sub-editor on the Bedford weekly paper, a debonair lad in his thirties who shared a flat on De Parys Avenue with his wife, a gorgeous nurse. Somehow Stanley managed to create in this sedate provincial avenue a life that would have seemed provocative even in the Latin Quarter of Paris, where I'd been hiding out for two years before surrendering to Her Majesty. Stanley had lived in London earlier, and drunk with Dylan Thomas, among many others, which is worth mentioning for the sake of one extraordinary anecdote. He told me that one night, at the George in Regent Street, there was a limerick contest, and Thomas had produced his own spontaneous contribution, which went:

The night that I slept with the Queen
She said, as I murmured *Ich Dien*
Darling do switch the light out,
This is royalty's night out,
The Queen may be had but not seen.

Of course I only had Stanley's word for it, but I never doubted its authenticity; it has just the right degree of wit, style and effrontery to fit with my notion of Thomas tossing off a line or two at the bar. Stanley wrote novels – or, rather, one novel, which was never finished. A lock of hair hung over his forehead and he had a sweetly insolent smile, devastatingly attractive to very young women, who buzzed around him like bees at the honey-pot. Almost every time I visited he told me with relish how his Lolitas were sabotaging each other's efforts, by deflating bicycle tyres, leaving false messages, getting out the police and generally plotting each other's downfall. His wife adored him, too, and tried to put up with it, but finally it was too much for her: she put his unfinished manuscript in the fire, and her head in the gas oven, from which he rescued her when he got home early.

He described later a moment of indecision – should he go first to the fire or to the oven? – but I am sure that was just a literary flourish, since otherwise the room would have exploded.

So that was De Parys Avenue in 1955, a street of big, gloomy, dilapidated houses full of flats and bed-sitters concealing who-knew-what goings-on, and all as shabby as the rest of England in those days. Fifty-five years later, it was a prosperous promenade, with manicured gardens lining the road, and rows of rehabilitated mansions stuffed with beds and breakfasts. But when I arrived the road was a clutter of sideshows and all the beds were taken, because that weekend more than 350,000 people were coming to Bedford for a festival I'd never heard of, to mess around on the banks and in the waters of the Great Ouse, a river I'd never heard of either.

My ignorance was moderately blissful. I had sneaked in on Friday, and things weren't going to get really riverine until the next day. My host was Mervyn Poulter, and without quite knowing why I decided that the name suited him very well. He was a distinguished elder statesman of the hospitality trade, having cut his teeth in the Caribbean, and he treated me well, but it was very clear that my time would be up on the morrow and, unless I wanted to sleep rough, I would have to move a long way away.

I wandered around Bedford and found it a much nicer town than anything I remembered. It was particularly pretty along the river where I may never even have been half a century earlier. The newer shopping area, a typical late-fifties orgy of cement, had taken a hit from the economic crisis and was looking rather bedraggled. It seems a shame to have to say, again and again, that the older buildings were the ones that still looked the best. I remember our infatuation with concrete at the end of the fifties, and I suppose it was understandable, it being such an easy lay, but we are repenting at leisure.

On my way back to the main street, I found what I think must be the world's longest pub and sank a couple of ales, and for once I managed to get into a conversation, with a honeymoon couple who told me that the river festival happens every two years and was begun in the seventies to celebrate a huge restoration effort. Apparently you can now navigate on it all the way to the North Sea, but I had to put off that pleasure for another time.

After my full English, served next morning by the impeccable Mervyn, I wandered off to see what was left of the place where I'd spent two quite tumultuous years. My return to RAF Cardington was only a duty call and I expected little of it, but I thought it would be fun to see those airship hangars again. I wanted to get a taste of the countryside too, so I took a circuitous route and lost my way for a while. The TomTom was not helpful because I had forgotten (or maybe never knew) that the RAF station was not actually at Cardington but closer to Shortstown, named after the Short brothers who built seaplanes between the wars and had had a lot to do with building the hangars.

When they did finally come into view they took my breath away. They are stupendous edifices, larger than anything one might expect. The bigger of the two is about eight hundred feet long and two hundred tall and, I now discover, could have contained *Titanic*, give or take fifteen yards. Now that I am writing about it I am astonished at how little I knew of what was around me when I lived on that RAF station, wandering about from hut to hut with my irons and my chitties. That whole airship saga continued for more than a century and is still alive today in spite of everything. There were spectacular disasters. The 1930 crash of the R101, which was built at Cardington, was one of them. The Americans had the *Akron*, which cost seventy-three lives off the New Jersey coast in 1933. The Germans had the *Hindenburg*, which ended in a fiery inferno in New York in 1937. Yet the dream is still a tantalising challenge and the hangars, an integral part of that dream, are monuments to the obsession.

Meanwhile people are always dreaming up new uses for them. Paul McCartney used one as a sound stage. The government had a Building Research Establishment there, the hangar being big enough to contain a multi-storey block that could then be deliberately destroyed or burned to the ground. In fact it was used to test conspiracy theories about what had happened to the World Trade Center on nine-eleven. Attempts to build new kinds of airship come and go.

Maintaining the hangars must be a significant expense, but they are as iconic and unique as the Parthenon or the Eiffel Tower. Imagine them in the centre of London: to take them down would be unthinkable. But they are tucked away in the country where few people find them and their future is uncertain. There should be a museum, with *son et lumière*, and holograms of airships bursting into flame, rides on blimps, indoor bungee jumping, souvenir shops, fleets of tourist buses ... Oh, my God, where am I going with this? Better leave them alone.

I came upon them this time from an unfamiliar side, over open country. A chain-link fence surrounded the site, and the locked gate carried a sign for the Driving Standards Agency. Apparently this was where people who administer driving tests go to be tested, and thinking about driving examiners cast a pall over my mood. I have never had much luck or sympathy with driving examiners. The one who failed me on my first motorcycle test in 1973 comes immediately to mind: a pedantic bald idiot who said I had stuck my knees out too far.

'I used to race at Brooklands,' he said, with waspish pomposity, 'and if I'd had my knees out like that I'd have broken them off,' which ranks among the most insufferable remarks ever made to me by anyone. I could rant on about it for hours.

A worried man, with a Middle Eastern look, drove up in an ice-cream van and asked if this was where his wife could take the driving test, but I was no help to him. I circled around the site for a while, finding nothing familiar, and thought I had better start worrying about a bed for the night. A look at the map revealed that not too far away, but far enough not to be troubled by the festival, was Milton Keynes. The name stirred up a strange medley of past thoughts and prejudices, none of them very favourable and all utterly unjustified, considering it was a place I had never seen.

Milton Keynes is a 'new town'. It was invented in 1967, although it turns out that there was already a village of that name, which scuppers one of my askance suspicions that it was a made-up name vaguely associated with the economist. It was a planned development of the kind we all got so excited about after the war. It was supposed to help with the housing problem, which, with unemployment, was one of the twin blights of pre-war Britain.

In my adolescence I used to believe in planning the way others believe in God. I had been brought up in wartime London so it was easy to assume that, since the government had won the war, it was obviously the new government's job to solve all our other social problems. They went at it with a will, piling up bricks and pouring cement over everything, and a lot of people did get homes, but so many of those quick and miraculous fixes had a sorry look about them later on. I'd always assumed that Milton Keynes would be another example of the same, and not my kind of town, but I decided to ride off bravely and give it a chance.

Bedfordshire is mostly arable land, growing wheat and other cereals. Pigs, I believe, do rather well there. It's pleasant enough but hardly dramatic and doesn't seem to have produced anyone of note, except perhaps Steve Linsdell, who specialised in almost winning the Manx Grand Prix on the most unlikely bikes, startling everyone with a second place in 1981 on a Royal Enfield Bullet that he had rebuilt himself. Anyway, nothing along the way made me reach for my camera. A little detour off the main road took me into Buckinghamshire and led me through Newport Pagnell, which struck me immediately as a rather pleasant little town with a nicely proportioned high street, but I ploughed on regardless, and a few minutes later I was in Milton Keynes. Well, I thought I was in Milton Keynes. I was never really sure.

I seemed to have found myself in a video game. I was on a rectangular grid riding around cells, and there was something weirdly pixelated and reddish-brown about the view. In the centre of each cell, a long way from the road, was a huge square block of buildings. From what I could see they were full of businesses and of no interest to me, but I imagined that at ground level there would be small, friendly shops, cafés and restaurants. If there were, I couldn't find them. I rode around lots of these cells, but never saw anything in the least attractive or interesting. I felt like a total failure. Somebody will tell me that I was in entirely the wrong part of town, but I made an honest effort and read all the signs, and nothing seemed to point to the right part.

The only other planned town I know is Chandigarh, in India, but by the time I got there Le Corbusier, who had designed it, would not have recognised it. The population had covered all his severe concrete with Indian signs, hangings, banners and graffiti, making it much more colourful and inter-esting, and I can't help saying that a similar treatment might have improved Milton Keynes. Later on when I examined the town's website I was just as confused. It rattles on about the vibrant atmosphere of the Theatre District, and it's hard to see how I could have missed it, but their map makes it no easier to find.

Apparently there are 350 restaurants, but on the map I can find only five. One of them, it says, is the largest Indian restau-rant in Europe. Is that a good thing? And there's something else I found alienating about Milton Keynes. It's an acronym. They call it MK, and seem to be proud of it. I could just about settle for an abbreviation, like Brum for Birmingham, say, but

I would not like to live in an acronym. I feel sure I am being terribly unfair to MK, but I fled, defeated, back to Newport Pagnell, hoping they didn't call it NP.

To punish myself for making such a mess of things I took the cheapest room I could find in my B&B book, and pretty soon I was outside a two-storey semi-detached brick house, just like the suburban house I was raised in seventy-five years ago. The other half of the semi was the Renji Centre for Chinese Medicine, offering Acupuncture, Acupressure and Herbal Medicine, and Dr Ling (or was it Ying?) answered the door, a slight man wearing a pale blue smock.

He looked at me in a quizzical way, as though my appearance was somehow inappropriate – I don't think either of us was quite prepared for the encounter. I admitted that I was, indeed, the one who had called about the room and, to lighten the atmosphere, I said that I was wondering whether it might be a good idea to have some acupuncture while I was there.

'What is wrong with you?' he asked politely. I fancied there was an edge of scepticism in his tone.

'Well, nothing really,' I said, and without another word, he took me up to my bedroom, which was also just like the bedroom I had slept in for the first five years of my life, with a very small bed, and windows overlooking the street, a womb with a view.

I went out again later, and the high street helped me to overcome my embarrassment at having failed the test of Milton Keynes. I walked up and down, asking myself again and again, why would anyone prefer MK to this? It may not be the most beautiful high street in the kingdom, but the orderly accessibility of everything was so obviously preferable. It was built on a human scale; it had all the right shops, selling cards and books, spectacles, coffee and groceries. There was an Indian, a Chinese, and four banks. The buildings spanned a decent chunk of British history and there was a very nice church at one end. Best of all, it had pubs, good ones that spilled out over the pavement on a lovely sunny evening. There was a place where you could see four pubs at the same time, with others around the corner. I wept for Wem and sat down at a wooden table outside to drink a pint.

Another honeymoon couple (what's going on here?), elegantly dressed and glowing with virtue, sat down across from me with glasses of champagne and, to my surprise, interviewed me about my life. This is something that never happens. He was handsome, she was lovely, they agreed with me about MK, and then left for some swanky dinner. I crossed the road to the Indian, ate very well and felt on top of the world.

The next day I made a life-changing discovery and, as with all such revelations, wondered what had taken me so long. I found that I could set my TomTom to a maximum of 30 m.p.h., and the result was breathtaking. It was like stepping back into a different world where everything was small-scale and local, where the important thing in life was not how to get to London or even the nearest big town: what really mattered was the shortest way for your horses to go from the hayfield to your barn. Most of the way from Newport Pagnell to Dorchester, I travelled on tiny lanes without names that skirted fields of rape and corn and cattle, sometimes between high hedges, sometimes with glorious views. I disappeared into forests, plunging through viridescent tunnels of greenery to burst out into sunlight again. Most of these roads were so narrow that two cars could scarcely squeeze past each other. Sometimes my path would dog-leg across a busier road, or pass through some small village, but hardly ever did I meet anything that could be called traffic; when I had to cross over a motorway or some other high-speed corridor I could look down and gloat over the poor creatures caught up in their mechanical madness.

This tracery of ancient pathways across the surface of Britain became a delight, and I wondered how a map of the country would look if only they were drawn on it, and all the major roads expunged. The idea became very pleasing and seemed to hold great significance. There are maps of the waterways, of course, and there are the railways, too, a pitiful remnant of what they used to be before Beeching truncated them, but still with at least one finger in every pie. And perhaps with hundreds of functioning airfields in the UK one could do something with them too. Different layers all revealing quite different aspects of the country.

Somewhere along the way I stayed in a room over a pub. The room was nothing remarkable, but it was comfortable and cheap, and the music downstairs was lively. There was a shower stall built into a corner and in the morning I got out of bed, naked as usual, and walked over to take a shower. Then I thought better of it, put on a shirt and glasses, and took another look. The controls were yet another arrangement new to me; not overly complicated but far from intuitive, and I would have to see what I was doing.

Anyone who travels much knows that water doesn't just come out of taps any more. It is channelled, mixed, regulated and dispensed by a huge range of sophisticated devices. There is some kind of primitive urge that drives us to invent ever more elaborate ways of tapping into it. I remember very clearly when I was a kid, and long before I'd ever seen anything more complicated than a tap, messing with it myself. The tap in our kitchen had come loose somewhere inside. I found that if I pressed down on it, water came out. When I let go it stopped. I spent some time thinking about how to make a tap that worked both ways: twist or press. Then I lost interest, like a million other kids, which is why we're not millionaires. But some did persist, and so today, when you enter a strange bathroom, there is always the thrill of encountering yet another baffling apparatus. Will you push or pull, twist, dial, slide or press buttons, or any combination of the above?

Of course, it's not just the problem of getting the water to come out: usually the controls are placed so that when you do get them to work, the first blast of icy water hits you in the face but, these days, I can generally work out a stratagem to avoid the dousing. That morning's shower challenge had the controls on the wall of the cabinet so I thought there was a good chance of success but, try hard as I might, no water came out, not a drop, neither hot nor cold. After I had tried every knob and lever in every combination I had to give up. There was no phone in the room, and even if there had been I don't know if I would have trusted myself to use it. Angry, frustrated and humiliated, I got dressed and went down to the bar, where a woman was serving breakfast. Through clenched teeth I muttered, 'What's wrong with the shower? I can't make it work.'

'I don't know, I'm sure,' she said. 'Did you pull the cord, then, dearie?'

'What cord?'

'You have to pull the cord first.'

'What cord? There's no cord!'

'Yes, there is, dearie,' she said. 'It's by the door.'

'By the shower door? There's no cord there.'

'No, dear, the room door. You have to pull it first.'

I wanted to say, 'Now pull the other one,' but I felt a growing conviction that she was serious, and that somehow, subtly, familiar things around me had changed, and my confidence that I could manage in twenty-first-century England was just a fantasy.

'Why is there a cord by the door?' I asked, with that hollow feeling you get when you know you're staring defeat in the face.

'I don't know, dear,' she said, 'but that's how it works. Now, I'm sorry, dear, but I must get on.'

Well, it is humiliating, of course, to be told that you have to accept as normal something that flies in the face of reason. Sometimes I yearn for the good old days before my time when, if you had the right voice, you could say things like, 'Now look here, my good woman . . .' or 'Do you know who I am?' and people would collapse into abject servitude, or so I'm told. I suppose it's a good thing they're gone, not least because I'd probably be on the wrong side of the dialogue. Anyway, I can take humiliation in my stride. I've been called the least threatening man on the planet, which is another way of ascribing to me a serial appetite for humiliation.

But this one I couldn't leave alone. I began questioning people relentlessly about their shower arrangements and learned that Britain now suffered under the tyranny of a domestic spying agency called the Health and Safety Executive, which spread its tentacles throughout the kingdom. With a little more prodding, I also discovered that there was a rumbling groundswell of revolt, particularly in the B&B community. Again and again I heard the words 'Health and Safety' uttered in contempt, and resolved to be on the lookout.

Even castles were defenceless against them. Later on my way west I visited Arundel Castle for the first time. This immense pile of stone and the gorgeous gardens around it are so wonderful that they are beyond my powers to describe, so I shall belittle them instead. For all their fortifications, they have been penetrated by Health and Safety.

At the foot of a towering keep is a grassy bank that sweeps down fairly steeply to the path. And planted in the middle of this steep bank is a sign that reads:

DANGER
STEEP BANK

I did make a serious effort later to expose the absurdity of the shower cord, and phoned Health and Safety about it, but they, of course, avoided the issue by referring me to some kind of electrical manufacturers' organisation, which in turn sent me to some academic association of electrical professors who were all on holiday in Portugal, so I didn't get very far. But I shall pin them down in the end.

12

An Ancient Mystery Resolved

As it happened, I'd read in an email that there was going to be an exhibition of some sort taking place at my old school in London, so I thought I would attend and see what it felt like to be back in that building. My fondest hope, though, was that someone would be there with some clue as to where I should be heading.

The school had long since closed, but the premises were being used by some sort of adult learning institution and still had that academic feel. Much of it seemed quite unfamiliar after almost sixty years. The asphalt playground and the bicycle sheds had been ripped away, and the space was enclosed by the high wooden hoardings of a building contractor. He was putting up a high-rise apartment block which would crowd the old school and, in my sentimental view, demean it. However, once inside I was fortunate. I knew that at one time I had gone west with a party of other boys to harvest potatoes. But where had we gone? There were a few old boys there, though much younger than me, and one of them said he'd heard something about those potato-picking jaunts. He was fairly certain that they'd taken place close to Dorchester, and had had something to do with the headmaster's wife, and another fellow remembered something about a village called Little Bredy.

I was delighted and surprised to discover that my dallying through the yesteryears of Britain would bring me back eventually to the CW Motorcycles showrooms. It seemed that the potato patch where I had once laboured would probably turn out to be only few miles from the town.

The first time I went west I was on a holiday to Cornwall, but I was too young, at five, to remember anything of the journey. Others I know have quite explicit memories from a much earlier age, but most of my early childhood is a blank. I must have been sent there by train because there was no other way I could have travelled that far, but with whom I went is a mystery. It certainly wasn't my mother. Sometimes, afterwards, I had dreams of flying low across the ground among trees, which I imagine must have been replays of my view through the train windows. One extraordinary image does stick, though, of a brightly lit train diving below ground in a big station, but I was never afterwards able to figure out where such a thing could have happened.

All the big stations fascinated me, with their immense, swooping steel and glass roofs, the noises of steam locomotives and echoing loudspeakers bouncing around inside those enormous spaces, the sense of a great world in motion, a thousand private lives glimpsed through windows in lit carriages, the rich bouquet of smells from smoke and coal, steam and cigars.

I remember only tiny fragments of the holiday itself. There was a beach where I tried to climb a steep stony bluff and became petrified halfway up, and a small cottage where I had an episode of yellow jaundice. I slept in a tiny bedroom where the walls were made of a creamy-looking plaster. There was no electricity, and I read Rafael Sabatini's *Captain Blood* by candlelight and first knew the sweetly acrid smell of burning wax from a guttering candle.

It was seven years later, towards the end of the war in 1944, the year of D Day, that I went with a party of kids on a mission to help win the war by bagging potatoes. Apparently there were many schemes to take schoolchildren on to farms to help with harvesting, but we were a bit special. Although ours was just a normal state grammar school, the wife of the headmaster, Guy Boas, was closely related to the Queen, and she had connections with landed gentry in the West Country who were persuaded to allow a party of grubby schoolboys to invade their property. We took a train (from Paddington, of course) and finished up in a field of potatoes that, touched by royalty, must have been the cream of the crop.

The potato-picking left little impression on my mind. All I remember is that we were camped in little canvas tents – bivouacs, they called them – and slept two to a tent on folding trestle beds. But there were diversions. On a seaside outing, which must have been to Weymouth, I found a bagatelle machine in an amusement arcade that took halfpenny coins. To my ineffable delight it was defective, and delivered rewards after every game. This came as a huge revelation. I still recall my unalloyed joy at discovering that there could, in this world, be free, unconditional gifts – until the arcade manager spotted me stacking up the halfpennies and rudely chased me away.

But what amazed us most of all was the hugely exciting military presence all around us. Up the hill there was a narrow road, and on the other side an immense dump of war matériel covered with camouflage netting, all destined to cross the Channel. American soldiers, generally of African origin, which added to their novelty since few of us had ever seen a black man, sauntered by in Jeeps, and were always ready to stop and distribute favours. We hung around like gypsies along the edge of the road, glorying in this opportunity to brighten our narrow wartime lives.

Usually what we got were cigarettes. At thirteen years of age my first smoking experiments (chestnut leaves rolled in newspaper) were still a few years off, but I knew the value of cigarettes. My mother smoked Craven A Cork Tipped when she could get them but often had to settle for anything she could find. Cigarettes were rationed and the most popular brands, Player's, Woodbines, Senior Service, were sold in austerity-paper packs of ten. Many newsagents sold them singly from cups on the counter, but often the shops didn't have any at all, and you would have to find a black-market shop that sold them for three times the price.

So, this glittering bounty of brightly coloured packs of twenties wrapped in cellophane was a treasure. Chesterfield, Old Gold, Lucky Strike, Camel, Philip Morris, this was our real harvest. We packed them away under our beds in big Huntley & Palmer biscuit tins, and competed to see who could bring in the most. But there were other things too. Chocolate bars, of course, for instant consumption. Even, occasionally, silk stockings 'for your mom'. We got tiny editions of popular literature, strangely attractive because they were bound in an unusual way, on the short side of the book and small enough to slide into a pocket – Charles Dickens, 'condensed for wartime reading'.

When the books, the fags and the sweets failed, the GIs handed out whatever they happened to have on hand, and that was how I came by an object that occupied my mind, on and off, for years. There were two tubes of some sort of material, about three feet long, joined together somehow side by side, which you could fill with air. I don't think I was familiar with the word 'inflatable', but that's what they were, with a solid brass hook at one end and a buckle at the other. The two ends clasped together perfectly, but it was much too tight to go round even our skinny bodies, and we were all baffled by it. Naturally enough my friends lost interest very quickly: it wasn't theirs and it seemed useless. For me, though, it had absorbing fascination, both for the puzzle of its function and its sheer quality. The brass work was tremendous – a quarter-inch thick with a lustrous golden glow. The chambers, which I suppose must have been rubberised, were made of an olive green fabric more finely woven than anything I had ever seen. It is difficult to express today the impact that pure quality made in those austere, utilitarian years. I kept the thing, as much for its beauty as for its mystery, in the little attic room where I had my chemistry set. Eventually, of course, I lost it in one of the inevitable upheavals of life, but was often reminded of it.

When I was in the RAF, dressed in a coarse woollen uniform, I met US airmen from a nearby air base who wore expensive-looking gabardine with the same close weave that brought it to mind again, and reinforced my notion of America as a land of careless luxury.

More than fifty years later, when I was already living in California, I heard someone praise the opening sequences of the film *Saving Private Ryan* as the ultimate description of the randomly murderous massacre of Omaha Beach. I got the film and watched it one evening. It was the arbitrariness of it that was chilling. Clearly there was nothing anyone on that beach could do to change the odds on survival. Whether you lived or died was a toss of the coin, and the mayhem was brilliantly portrayed, but I remember feeling cheated because the presence of Tom Hanks, who was obviously destined to survive, undermined the authenticity.

Then, in the aftermath of the slaughter, as the camera patrolled lazily over the wreckage and examined the bodies strewn over the beach and in the water, I saw my long-lost object. I had not thought of it in decades, but I knew it immediately. It was wrapped around an ammunition canister, bobbing gently in the waves, and the shock of recognition was profound. To have a mystery resolved so long after and so accidentally had a touch of romance and magic about it, as though I had stumbled on a letter that made sense of some inexplicable turn of events in my distant past. It took me instantly to that time and place where some unknown American soldier, no doubt long since dead, had planted this seed in my life.

I have more experience than most of returning to places that were significant decades before, and I had no expectations of a *eureka* moment, but it all fitted rather well into my general scheme and I knew that the effort of locating the place where I'd come by my curious artefact would drag forgotten details from my mind's attic to warm up and refurbish my memories.

13

Life on the 31 Bus

Of all the roads I travelled in my youth, by far the most important was the one that connected Swiss Cottage in the north of
London to Fulham Road near the Thames. Of course, as it ran
through north London, it wasn't just one road, but many –
twelve or more, depending on how you count them – but they
were all unified by one common denominator: they were on
the route of the 31 bus. This was the axis of my life for seven
years, from the age of twelve to when I went to college.

At the southern end, a quarter of a mile from the bus stop at
the foot of Redcliffe Gardens, was my school – a quarter of a
mile I often had to run in desperation to avoid ignominious dis-
approval or worse. At the northern end of the bus route lived
my playmate, Stephanie, who became the girl I wished I had
the courage to call my girlfriend. I lived in the middle of the
route, at Notting Hill Gate. Every weekday morning, just after

eight, I raced to the bus stop (I can never remember having enough time to walk), and willed the bus to arrive NOW.

When it did, I swung on to the platform and rushed upstairs, hoping for an empty seat at the front, where the conductor would eventually find me. He carried a little rack strapped to his chest, and on this rack were six or seven stacks of tickets held in place by a spring-loaded lever. The tickets were about one inch by two inches of rough paper, and printed in various rather nice candy colours, depending on their value. Across the top of each was a serial number, and below that the route number and a list of the major stops, so bus tickets issued for the 31 bus route couldn't be used on the 28, for example. I don't think this was a security measure: it was just that we still lived in an age where bus routes and prices seemed to be fixed for eternity. You could print as many tickets as you liked for the 31 bus, because sooner or later they would all be used. My journey to school took about half an hour and cost a penny. Next to his ticket rack the conductor had a metal punch over a leather pad, and when he sold me my ticket he punched a hole next to the words Fulham Road or Notting Hill Gate, and the punch went 'ding'.

Inevitably buses and their tickets all played a part in the fantasy world of childhood. The two-decker London bus was a formidable machine and each one had its own number on the bonnet. You wrote them down in a little book, and there was respect to be gained by spotting more bus numbers than your friends. There were mysteries, too, which I have forgotten, associated with the serial numbers on the tickets. And because the buses in those days all had open platforms, with a single vertical rail to hold on to, there were challenges to be met when running to catch a bus or jumping off at the lights – all much too dangerous, of course, for our modern, foolproof age. It was generally assumed that people would behave reasonably, and the conductor was there to control things, with shouts of 'Move along, please,' and 'No standing upstairs'.

Although I never personally saw an accident caused by letting passengers take their lives into their own hands, there must have been some. My cold-blooded inclination is to believe that there are people who will find a way to have an accident however hard you work to prevent it, and the more you do to prevent it, the more tedious life becomes for the rest of us. Perhaps this cynical attitude was born during those days on the 31 bus because quite often, as I rode to school, a house that had been there the day before was now just a heap of tangled iron and masonry where, during the night, a flying bomb had landed or a rocket had ploughed into the ground. Coming down Church Street to Kensington High Street one morning, I passed a block of flats where one floor had been completely blasted out, save for the girders, while the other floors were apparently untouched. Another day, on the same road, another building had had one wall quite stripped away, revealing a kitchen with the table still set for breakfast. And then, at the corner of the high street and Earl's Court Road, there was a big restaurant that was blown up during the lunch hour. Nobody could tell where the next bomb would fall. Life was evidently a lottery.

The war intruded on our studies quite a lot, directly and indirectly. The school was more crowded than it would have been, and our *nom de guerre* was the West London Emergency School for Boys, but thirty to a class was not intolerable. We were divided into four streams, A, B, C and D, which rarely mixed, unless there was an air-raid siren when we were hustled into the library, which was considered safer, being in the centre of the four-storey building. This library became like a Middle Eastern bazaar, with all of us trading copies of *Hotspur*, *Wizard* and *Champion*, prize conkers, stamps and coins, but all this ended soon enough: the bombs began coming so erratically through the day that the sirens became obsolete and we would have had to spend our entire day in the library.

Another interesting consequence of the war was that our teachers made for a much more colourful crew than the normal peace-time establishment: many had been dragged out of retirement to replace the young men who had gone to fight. We also picked up a few strays, who might not otherwise have fitted the image that our school was known for, because Sloane School had had something of a reputation before the war when our theatrical productions were reviewed in the press.

One of these round pegs in our square hole was a young woman from New Zealand. Heaven knows what her name was, or what she taught, and I don't think she was with us for very long, but she was pleasant enough, if a bit odd. She wore a lot of jewellery, and dresses that were a little advanced for our austere times. One day, during a period when the buzz-bombs were coming over more frequently than usual, we were assembled in one of the larger classrooms with a high ceiling and long vertical windows, obviously susceptible to blast, so I suppose we had reason to feel a little nervous. She sat in front of us, on a dais behind one of those large wooden desks that have drawers down each side and nothing in the middle – they were

sometimes useful for providing an accidental glimpse of knickers, although not on this occasion.

'Now, I know you might all be a little anxious,' she said, 'so
I want to tell you that you have nothing to worry about. You
see, I have a very special necklace . . . ' and she took a string of
stones from around her neck ' . . . given to me by a Maori chief.
Now, if danger is looming the necklace will tell me in plenty of
time, and I will tell you all to get down under your desks.'

It could not have been more than five minutes later when a
flying bomb landed nearby. There was an almighty explosion
close enough to rattle all the windows and distract our attention. When we looked back our teacher had disappeared, only
to be discerned cowering under her desk. I can't remember
seeing her after that.

Another cranky old teacher made an indelible mark on me,
and taught me a vital lesson, although it was not the lesson he
thought he was delivering. He was a maths master called Nightingale, an extraordinary scarecrow of a man, tall, lean and bony,
whose tattered gown flapped around him (for our teachers still
wore those black gowns in an effort to preserve standards). As a
teacher of geometry I must say he was excellent, and it was easy
to forgive his eccentricities, such as tearing handfuls of cloth off
his gown to wipe the chalk off the blackboard.

It was towards the end of the war that he revealed another
facet of his personality. The Allies had finally forced their way
into Germany, to discover the first of the concentration camps
at Buchenwald and Bergen-Belsen, and the papers were full of
pictures of the walking and recumbent skeletons in striped
prison garb lined up against the barbed wire. Their gaunt and
hollow faces were the stuff of nightmares.

Nightingale strode up on his dais and gazed fiercely at us. 'Don't believe everything you're told,' he cried. 'It's all propaganda. I can look like a Belsen victim if I want.' And Nightingale, who was pretty gaunt to begin with, sucked in his cheeks and did a very dramatic imitation of one of those dying men. We in Form 4A were not of an age to be particularly offended by this piece of extravagant theatre. I don't suppose many of us thirty-odd fourteen-year-olds even knew what he was talking about or why. There were glances of amazement and then, as they used to say in the music hall, 'A titter ran around the audience.'

Because of my mother's activism I was already something of a political junkie myself. I had a rough idea what he was getting at and I thought he was nuts. I knew that during the First World War there had been cartoons showing 'The Boche' with babies skewered on their bayonets, and, thinking about it later, I guessed that maybe Nightingale had been one of those Englishmen who had found a way to admire the Germans before the war. But I already had some idea of what the Nazis might be up to, and in any case, the evidence in the papers and newsreels was overwhelming.

One can easily imagine what would happen to a teacher who came up with a performance like that today. The very idea of a teacher expressing eccentric personal opinions to a class of 'impressionable minds' would raise the concerted hackles of a thousand education officials. He'd be hustled off to rehab or retirement before the day was out. It's a terrible shame. Our minds may have been impressionable, but they were a lot tougher and more resilient than one might imagine. I don't suppose for a moment that any of us believed what he was telling us. The lesson I learned from Nightingale was that a man can be nuts in some respects and still do a great job or be a good person in others.

Where I live, in the backwoods of California, I maintain friendships with creationists. I think they're just as kooky as Nightingale, but one of them is a judge, who could be counted on to give me a fair shake when I fell foul of an American law I had never heard of – the infamous 'open container' law. In South Africa in the seventies I was helped out of trouble by a very kind and pleasant man who believed without question that blacks were inferior to whites. A friendly Arab peasant in Tunisia invited me to tea, looked me straight in the eyes, told me how terrible the Jews were and that he could smell one a mile off (well, I'm only half-Jewish; maybe that explains it). Everyone has stories like this to tell. A very important part of growing up is learning how to find common ground with others and to cope with their eccentricities later, from a position of goodwill. Unwittingly Nightingale helped to build that useful resource. But today no hint of contention, no unpopular viewpoint must ever sully the bland atmosphere of the classroom. Without controversy education is boring, and the best teachers are often excluded.

There were some occasions when I took the bus after school all the way to Belsize Park where my friend Steffi lived; a long journey that could take an hour or more. I couldn't have been older than eleven when we first met and played as children, but we continued to see each other for six years, through all the years of puberty, until the relationship came to a humiliating climax.

I suppose it must have been my mother who first met her

parents, again through the refugee network. I think they were from Czechoslovakia, but it might have been Austria. I have a vague impression of him as a somewhat rotund, balding man, a jeweller who worked in a back room, but I have no recollection of the mother at all. I suppose they were only too pleased to have someone to keep Steffi occupied, and left us alone. It was two or three years before I began to appreciate that she was actually a girl, with emerging physical characteristics, but because she was the only one I knew, and we had played as children for so long, I could never find a way to cross into that quite different, tantalising but frightening world of emotion. Burned into my memory is the day in those in-between years when we were in a park with small trees and Steffi, for reasons that even she probably couldn't quite identify, decided to climb a tree above me, and I got my first ever glimpse of a female pubic hair. Even that was not enough to push me across the line.

The *dénouement* came when we were both sixteen and I persuaded her to come to a dance at the school. By then she was a ravishing young woman and I was adoring, but didn't have the guts to do anything about it. We wandered about the school between dances, visiting classrooms and encountering couples snogging. Thinking I might play the piano for her, I took her to the music room, a little eyrie at the top of the stairs, but a school friend was already there, playing very fluent jazz piano, which did nothing for my confidence. Finally, on our way downstairs, she lashed out at me scornfully, 'Why didn't you take me in your arms?' So I stretched out my arms and she said, 'It's no good now. It's too late.' And stormed off. And that was the end of that.

It was more than a year before I finally overcame my shyness, but that's another story with another person. Steffi and I met again some thirty years later; she was a dancer and still beautiful, but the timing was wrong.

Now that I'm thinking about my childhood in the streets of London I am full of wonder at the mystery of what I have become; how I got here from there, from Notting Hill Gate to the Wild West. There were clues, of course, certain moments I can recall, thoughts that passed through my mind during my last years at school, but they suggested nothing more than a certain disquiet, a sense of unreality. Oddly they seemed to come mainly when I was walking along a street.

One of those thoughts was that I simply could not imagine myself as someone who had successfully gone through university and come out at the other end with a degree and a profession. The idea just didn't seem possible for the person I was. And yet there was no question of my commitment. I was passionately interested in chemistry. I had my own little laboratory at home in what used to be a scullery in the basement. I had a lot of complicated glass apparatus, a microscope, a balance in a case from Griffin & Tatlock, and a shelf full of chemicals, some of them quite lethal. And these weren't just toys. I used them. I knew a lot about building organic compounds from basic ingredients, and I had already caused an explosion in the kitchen, which prompted a cat to jump through a three-inch gap in a window six feet off the ground.

My life in the late forties in London was dominated by chemistry, politics and girls – or, rather, the lack of girls. I couldn't get it on with Steffi, and the time did not favour chance meetings. It wasn't so much that sexual activity was frowned upon as that there were few environments to foster it and few practical opportunities to practise it in private. There were no clubs, coffee houses or discos to raise the emotional temperature. Pubs were painfully middle-aged and dreary. Most homes were crowded. For the more confident among us, of course, there were no obstacles: any park railing or wood-shed would do. How I envied them. How did they know which girls would and which wouldn't? How did they know that sex would triumph over conversation, that politeness and a mutual interest in literature and philosophy were not likely to bring about the desired result? To my enormous relief, I finally lost my virginity, along with ten shillings, to a young prostitute on Bayswater Road who ensured my protection and hers with a French letter of industrial strength. She was kind enough to wonder why some nice young girl hadn't wanted to accept it as a gift, and her words were sweetly encouraging.

And then there were politics. For two or three years I was absolutely convinced that Communism was the only intelligent solution to the world's problems. I canvassed at elections, I spoke at meetings, I badgered my friends. Although I was a quivering wreck when it came to confessing my feelings to girls, I was a brazen lion addressing complete strangers in the street and touting the Party's literature (and it does seem ironic that 'literature' was the word we used for those polemics and manifestos that I thrust at people). In no way do I regret those years. My motives were utterly pure and my activities did no harm. I also learned that, provided I wasn't pursuing my own self-interest, there was no limit to my social bravery. But it has led me to wonder whether compensating for sexual frustration might not be a major component in driving young people today to extreme positions, including suicidal acts of terror. After all, they do seem to come most often from societies or communities where the repression of sex is also extreme.

I was nineteen before I began to doubt that human beings were capable of the kind of idealism a system like Communism required. In Paris I read Arthur Koestler's *The God that Failed* and that helped to usher me into a political no man's land that lasted for a decade or more. It was a strange and uncomfortable position to be in for someone who was used to arguing vociferously for a particular point of view. Where before I always had an opinion or an explanation, I now floated rudderless in a sea of ideas. I felt literally at a loss, and I understood why renegades like myself rushed so often to the opposite pole, of the Tory Right, or the Catholic Church, or both; anything to fill the void. What enabled me to resist was my rather peripatetic life, which exposed me to other ideas about meaning and purpose and my own place in the world.

14

Do You Mind if I Dunk?

As I followed again that long, rambling route from Fulham Road to Swiss Cottage, it appeared to have changed remarkably little in sixty-five years. For much of the way it passes among large older houses, in Redcliffe Gardens, Pembridge Villas, Chepstow Road and Belsize Park, or mansion blocks on Earl's Court Road and Church Street. These buildings may well have been rebuilt on the inside but externally they remain much the same. The big department stores on Kensington High Street, Barkers, Derry & Toms and Pontings, were once the height of sophistication. Derry & Toms had a roof garden which still survives, although they have all passed through several adaptations, and the rather curious former town hall facing them, where I remember once going to 'Saturday Night Hops', appears unchanged.

Notting Hill Gate has, for the most part, been rebuilt but I have a good recollection of how it was, thanks to a dog I owned in the late forties. Curly was a black and white mongrel with wiry hair, rather like a terrier, I suppose, and I liked him a lot for his gravity. He had a very independent nature: he left the house in the morning and came home in the late afternoon. One day I thought I'd spy on him to see where he went, and I discovered that he had actually claimed a huge territory. First he wandered into a churchyard opposite, where there were some plants he nibbled at, then he sauntered off up Kensington Park Road and over the hill past St Stephen's and down to Portobello Road market – just a vegetable market in those days – where he had a fight with a black Yorkshire, and collected a scrap of meat from a kind-hearted butcher. After he'd wandered on, I learned that this was a daily occurrence. Then he went back up the posh end of Portobello Road, where the residents had been clamouring unsuccessfully to have the council change the name of their part of the road to dissociate it from the riff-raff down below.

When he got back up to the Gate there was another butcher who favoured him with a daily tid-bit and then, to my amazement, he negotiated all the dense traffic of Bayswater Road with aplomb and crossed into Kensington Gardens, finishing his day with Peter Pan and the Round Pond.

The butcher, the pub next door and the other little shops along that side of the road were all swept away by an office development. Similar massacres occurred on the south side of Bayswater Road, leaving only the cinemas intact but wiping away the first post-war coffee house of that era. Desecration struck across the road around my morning bus stop too. The tube station vanished, and the ABC tea shop round the corner is particularly missed because it was the scene of one of my most surreal episodes. I had gone in for a bun and a cup of tea one frosty morning when I was joined at my table by an older woman wrapped up in scarf and overcoat. She sat opposite me and peered out with rheumy eyes from beneath a woollen concoction on her head. Then she leaned over slightly and confided, 'It's awfully foggy at Charing Cross.' I nodded politely and made sympathetic noises. There was a moment's silence while she chewed on her bun, and I completely failed to notice that she didn't have any tea. Then she leaned forward again and said, 'It's awfully foggy at Charing Cross.'

Well, I was a little unsettled by this, so when she quickly added, 'Do you mind if I dunk?' I said something like, 'Of course, go ahead,' at which point she leaned right over and dunked her bun in my tea. I have no idea what I did next.

On the way to Swiss Cottage the 31 bus crosses Abbey Road, with its recording studios where the Beatles did that famous cover shot for their album of the same name. I never met the Beatles (well, there's a startling admission!) but I did get closer than most: I was in the basement of their Savile Row building with their manager, waiting for them to come down from a session, but it was the day they decided to break up and they never emerged.

I've had bad luck with rock groups in general. I was on the *Daily Sketch* when Andrew Oldham phoned me to ask if I'd be interested in doing something with this new group he'd found called the Rolling Stones. I said to myself, 'Oh, God, not another rock band,' and I turned him down. But I did once get to watch Manfred Mann make a recording at the Abbey Road place. They were doing a song that was never released (as far as

I know) but I really liked it, and I can remember it to this day. It began, 'The world is just the world Apollinaire, I floated out into the market square . . . ' and the images were quite surreal.

At the end of the 31 bus route, Swiss Cottage has become a whirling doughnut of traffic with no discernible appeal, but the most interesting changes for me have taken place at Westbourne Park. This is where the road crosses the main-line railway from Paddington and then – to my surprise – crosses another bridge over the Grand Union Canal; surprise, because in all my childhood years I don't ever recall noticing the canal.

What I did notice – and I have held on to it ever since – was an abandoned cinema, an Odeon, I think, with the letters of the final attraction jumbled on the marquee. They read:

GNEEN BEEFTOT SO

What made this absurd jumble so fascinating I don't know. There are a thousand things I've forgotten that I wish I could remember, but this ridiculous riddle – GNEEN BEEFTOT SO – I cannot forget. I imagine that at the time I was so involved in trying to reconstruct what those letters once said that I missed the canal. Today that cinema has gone without a trace. There is a small group of buildings in burnt-umber brick among mature trees that must have replaced it, though they appear to have been there for ever.

I stopped and looked over the parapet of the iron bridge, painted a bright glossy blue and outlined in a shade of tanger-ine, as if to illustrate the difference between the two worlds, above and below. A notice told me it was the Grand Union Canal. On the north side was a terrace with chairs and tables belonging to a pub, which must also have been there in my childhood. I went in, bought myself half a pint of bitter and a bowl of pea soup, and sat by the water thinking about canals.

One evening after dark, when I was driving with a friend through the Hammersmith area, two cars passed us, one following the other in close succession. They were both black and unmarked, and in the second or two that I saw them I thought they might have been Jaguars. They were moving at a speed that seemed utterly impossible. There was a reasonable amount of traffic on the road but they moved through it silently, effortlessly, and so unbelievably fast it was as though we were at a standstill. Indeed that was the effect they had on me. One second they were there, the next second they were gone, and we had hardly moved. Their drivers' ability to manoeuvre through the traffic was superhuman. They might have existed in a different dimension. Never before or since have I seen anything like it. I waited to hear the crash which must surely come, but they vanished into the night without a sound. Who and what they were I never knew. Was one chasing the other? Were they together? Nothing surfaced in the news, and there was no one to ask.

The memory of it recurred many times. Gradually it settled into a realisation that while we follow our normal behaviour and assume it to be the only reality, there may be others living at a quite different level which we only occasionally glimpse. A similar window opened when I once saw, quite incontrovertibly, a ghost – or rather, as would-be experts later told me, an 'astral projection'. It changed everything about my beliefs, and nothing in my life.

In a more prosaic way, perhaps, I suddenly became aware, looking down at the canal, of an entirely new landscape of

Britain and a completely different way of life that existed among us but virtually out of sight.

The strange thing is that I have always known about canals but never fully *realised* them. A long time ago I learned that they were built as the Industrial Revolution got under way in the late eighteenth century, and that they were overtaken and made largely redundant by the sudden and dramatic growth of the railways. In my wanderings around the roads of Britain I would occasionally come across them, in Gloucester, in Manchester, in Bedford and Bristol, but I had never made the positive effort to connect all these sightings, and grasp the relationship between them.

If necessary, I can call up in my mind's eye a rough picture of the shape of Britain's road network, or at any rate the major roads and motorways, and how they radiate from London, but I have no sense at all of what the canal network would look like or how extensive it might be. All of a sudden, sitting by the canal at Westbourne Park Road, I had the feeling that I had missed completely an entire dimension of British life, and the prospect of discovering it seemed exciting.

I scraped together a little history of my encounters with canals. I knew from my visits to Lois and Austin how peaceful and remote life on a canal boat can be. Theirs is moored not far from Heathrow, on an arm of the Grand Union. The accommodation is far from luxurious, but it is cosy, there's room for a guest or two, and there's the simple fact that, with so little space, there are necessarily far fewer things to worry about, which is a great advantage to some and totally unacceptable to others. Indeed, if we all lived on canal boats the consumer economy would clearly collapse.

When I lived on Upper Street in Islington I walked a couple of times on the towpath of the canal behind the Angel. There were brackets on the wall at regular intervals that were meant to hold life belts, but the belts were missing. Instead there was a notice from the council saying that the belts had been stolen by vandals, and ending 'We apologise for any inconvenience', which I found darkly hilarious. As you drown, someone on the towpath is muttering, 'How inconvenient.'

Then there was my friend Eliot, and his wife Meg, whom I met in Kathmandu in 1977, and who now live near Stroud. I visit them as often as I can and they have taken me twice to see the Sapperton Tunnel, which begins only a short walk from their bit of land. At first it's hard to appreciate that there's a canal there at all: it looks like an overgrown creek bed where water might run in the winter. The opening to the tunnel is like one side of an old bridge and looks like a relic of ancient masonry lost in the woods, but this tunnel was one of the engineering wonders of the world when it was finished more than two hundred years ago in 1789. It is two and a quarter miles long, and so straight that when it was functioning you could look through one end and see daylight at the other. It took five years to build and it consumed a good number of lives in the process. It was, for a while, the longest tunnel in the world.

Obviously a project as costly and ambitious as this would have been undertaken only if it served a really important purpose and could be expected to make or save a mint of money. In fact, it was an essential link between the Thames in the east and the Severn in the west, between London and Bristol, a waterway which, but for the tunnel that acts as a land bridge, would have split southern England off as a separate island.

The tunnel itself is in disrepair and here and there it has fallen in, but there are plans to restore it. In fact, a tremendous amount of work has been going on for years to clean up and revive the canals because they provide important opportunities for tourism and recreation. Around Stroud itself we saw major engineering works in progress to reclaim the parts of the canal that had been built over or filled in.

Nicholson's map of waterways offers a fascinating and quite unfamiliar view of Britain. Aside from revealing just how widespread the canal system is, it is also naturally a map of watersheds, so that you can see at glance where is up and where is down. Salisbury and Winchester are up, as befits cathedrals, and so, I'm afraid, is my nemesis, Royal Tunbridge Wells. Grantham, Shrewsbury, Mansfield and Melton Mowbray are all up, while the big cities, usually built beside rivers, are down, although Bradford, Leeds and Sheffield seem to be more up than most. Well, aside from this child's view of geography, what the map does very well for an innocent like me is offer an enticing prospect of being able to putter around Britain at four miles an hour, far from the howl of motoring mania. Here is a network of waterways that actually represents a complete alternative world, and with my growing abhorrence of cars and motorways, the thought of the calm quiet of the canals is tremendously seductive.

I trained my Google eye to find Lois's boat, where I'd spent such pleasant nights tucked away on the edge of London, and out of curiosity I began to follow her canal, the Grand Union, north on a simulated cruise, tracing its progress through Uxbridge, Rickmansworth, Watford, Berkhamsted, Tring, and Leighton Buzzard. As I went I became enthralled by the immensity of the project, miles and miles and miles of tree-lined waterway passing through towns, under bridges, past the end of some lucky person's garden, behind industrial estates, occasionally alongside roads, but always shy and reticent, hiding away behind its bosky fringe. I noticed with approval the large number of pubs and restaurants situated alongside the water, providing sustenance and good cheer (I can't abandon my myth) at the end of each day. I followed it round the edge of my other old nemesis, Milton Keynes (MK for short), and then somewhere, but I couldn't find where, it crosses over or under or through the Great Ouse, a river that at this point is not yet great and has a spectacular series of loops to perform before it achieves greatness in Bedford, where I first met it.

It is immediately obvious that digging the first canals must have involved much boldness and huge resources. After all, a bit of road can be used but a bit of canal is useless. It has to be finished to get anywhere, so it's not surprising that the first canals were built by extremely wealthy land-owning aristocrats, like the Duke of Bridgewater, who financed the first canal to take coal from his mines to Manchester. It ran partly underground, it crossed a river on an aqueduct, and it cut the cost of coal in Manchester by more than half. The success of this enterprise led to an astounding burst of activity: within a century more than four thousand miles of canals had been dug, and most of them are still there. I decided to put this new-found vision of a silent Britain to the test, so I cornered Lois in a café near Paddington. She arrived on a motorcycle with a ukulele strapped to her back, halfway to an office in Islington where she works. Her work consists of writing a book. I thought I'd make the point, once more, that writing is actually work, although she claims to enjoy it. Anyway, I told her about my discovery at Westbourne Park.

'Do you know that pub?' I asked.

'Yes, yes, very well. I've been up and down that canal countless times. It's called the Paddington arm. It comes off the main one we're on. It goes right past Kensal Green cemetery. Do you know . . . ?'

'Well, no, I know nothing about anything you can't see from the thirty-one bus. Are there pubs all the way along?'

'Yeah, quite a few. Where there's a bridge there tends to be a pub.'

'So the one that goes through Islington, that's the same one, isn't it?'

'Well, it seems like the same, but technically it's the Regent Canal. They were owned by different canal companies, so you had to pay a toll. And there's a toll house at Little Venice, for example. That's the really fancy place to live on a boat.'

'But if Islington is basically on the same canal as your boat,' I protested, 'why don't you go to work by boat?'

She laughed. 'From my boat to my office would take a long day. It's not so bad until you go over the North Circular . . .'

'Over it?'

'Yes, you can see the boats crossing it. It's lovely looking down and seeing the cars all stuck in a jam. But after that it gets very slow.'

The boat she lives on now with her husband is not her first.

'I bought my first boat in Nantwich, in Cheshire. It was on the Shropshire Union canal and I took it all the way down and then round by the Coventry canal, picked up the Grand Union and came down to London that way. The brilliant thing about the waterways in Britain is that there's no driving test you have to take, no regulation. You can literally buy a boat and set off. No one to stop you. Really refreshing.

'That was a very exciting journey. I didn't know what was happening – didn't have a clue. The locks are probably where things can go wrong, but if you do bang into something you're only doing four miles an hour. And you can stop anywhere you like. The official rule is you can stay anywhere for two weeks, but people often stay longer.'

I was remembering my virtual tour of the Grand Union, and all those long, narrow objects, like stick insects, clinging to the sides of the canal or bunched up in lay-bys. Even on the relatively short distance I'd followed it there must have been hundreds. Lois is the only canal dweller I can recall meeting, yet I imagine that she's part of a population that must number in the thousands.

'It's really a secret world,' she said.

15

In Memory of Lauri

My long and beautiful ride West took me through Sussex, where I was reunited with two old friends I hadn't seen for more years than even a donkey could remember. One of them, Lesley-ann, was an old flame, almost as old as I am, who inhabits a splendid, rambling old house in Brighton, but pretends to live in a shed at the bottom of a bowery garden. She told me, to my surprise, that she remembered me as being a kind person, and showed me aspects of her town I had never known.

The other, Lauri Say, was the only old school friend still alive, as far as I knew. His life had taken an almost diametrically opposite course to mine. His first marriage, when he was very young, was to a cousin of mine, and they lived on the Isle of Wight. Later he got married again, to Marie, and moved to the mainland, but he stayed in the same part of England all his life – if he'd travelled at all I never heard of it – and had numerous children. At school he'd been one of the funniest people I'd ever met, with a broad, happy face, and a great appetite for music of all sorts. He taught for a living, engaged noisily in local politics, and wrote and sang songs in troubadour fashion, which enlivened thousands in the general Portsmouth area. He reached the pinnacle of his public life, probably, when he helped organise the British equivalent of Woodstock on the Isle of Wight. We hadn't met in something like sixty years and I was very lucky to find him when I did. He was bedridden as the result of a terrible car crash some years earlier, and was fighting valiantly against a deep-seated infection that defied medical science – if it is a science. We had a wonderful conversation. And then, just a few months later, he died.

I arrived in Dorchester as to an old friend but had to make do with a room at the George. David Wyndham was on holiday, and the Beggars Knap was fully booked. I had a bit of work to do at the post office with passports and licences and then took my time looking around the town. I found a very enjoyable restaurant with an open garden area, reflected sombrely on the fate of the victims of Judge Jeffreys, the 'Hanging Judge', whose rooms are still maintained, and found other interesting nooks and crannies, but the best time I spent was by the River Frome. It's only a small river running through the town, with a path alongside it, but there is plenty of wild vegetation on both banks, cow parsley, nettles, ivy and so on. I sat on a bench, watching the water ripple past peacefully, and a large, vigorous woman passed in front of me, carrying a lead. At least fifty feet behind her, a minuscule dog was trotting very purposefully, at exactly the same pace, not in the least concerned about the distance from its mistress. The two of them formed a procession. Absurdly I thought they should be on the river, a big barge and a tiny tender, both moving at the speed of the water. I tend to forget that Britain is a country of rivers and the thought crossed my mind again that the waterways of Britain should have some part in my exploration.

At CW I asked Wally if he knew much about what had been going on in the area during the war, and he had some stories to tell, but they had mainly to do with the RAF and various bits of dodgy business.

'They were bringing cans of petrol down here and selling it. And that was the officers. God knows what the men were up to. But if anyone knows about Little Bredy, it'll be Richard, but he's not here today.'

So I set off in the morning. It was only a few miles. I could come back and talk with Richard any time.

Among the most productive excitements of travel are those occasions when you come upon something so completely unexpected, so much out of scale and context, that it shakes you loose from your moorings: the glittering palace in the jungle, the elephant in a suburban garden, a painted warrior emerging suddenly at the roadside. Coming unprepared upon those Cardington hangars could really startle a person ignorant of airships. Something similar happened to me after I left Dorchester.

The land around me undulated in folds, concealing and revealing views as I rolled along. I thought I had left the city well behind and was riding through open country when over the hilltop on my left suddenly appeared a row of structures that had no business being there. They looked like the tops of large buildings and I remember my first thought was how insubstantial they seemed, like a film set erected the day before. Then the hillside gradually levelled off, the road veered around it, and I found myself looking down a long street, unusually broad and straight, just like the main street of a western cow town. But this was no film set.

The rooftops I had seen belonged to a tall modern block at the head of the street, but after that the houses on either side were solid examples of prosperous British city dwellings of the eighteenth and nineteenth centuries, generously separated, evidently occupied, with curtains in the windows and well-tended gardens. A little way along on the right, a middle-aged couple, of the kind you would see played by actors in an advertisement, were unloading things from the hatchback of their vehicle.

Yet there was something about it all that cried out, 'Fake!' Unable to say exactly why, I knew this could never have come about naturally. The road was wrong. The geometrical straightness of it, the disproportionate width of it, the complete absence of parked cars and the spacing of the houses alongside it looked wrong. For a little while, schooled by countless spy dramas and science-fiction tales of parallel universes, I felt I was the victim of some incomprehensible conspiracy. Was it a Potemkin village, perhaps, designed to delude terrorists? At the far distant end of the street I could see now that there were tall cranes and hoardings. My goodness, they were still building it!

I went across to talk to the humans and learned the truth. I had simply never heard of Poundbury. My humans, who turned out to be very real, pleasant and amenable, told me what most readers will already know: under the patronage and with the close attention of the Prince of Wales, a new town had been built – was actually still being built – alongside Dorchester, and the idea was, I believe, to incorporate the best of what had already been achieved in recent centuries while avoiding the worst of what had happened organically, like, for example, streets choked with parked cars. By sheer chance I had stumbled upon the antithesis of MK, and through even greater luck, I had a chance to see it before I knew what it was.

My problem was, I didn't know what to make of it. I am not automatically a fan of progress. I like electronics and high-speed rail, but I like living in old houses. Because there are no old houses where I live, I have been building my own old house for the last seven years, and, yes, I am trying to do pretty much what Prince Charles had in mind. It doesn't look like an old house from the outside: it can't, it's supposed to look like a barn. But inside I am using all the old ideas that I liked from houses I have lived in and visited in the past. I prefer rooms, as opposed to having everything in one large open space. I like moulded wood and plaster, picture rails and tiles, high ceilings, and windowsills deep enough to hold vases of flowers.

Because I am taking so long over it, the house keeps changing as it grows and, to tell the truth, I'm sure that most of the friends who visit think it's just a mess. It's definitely a work in progress, and a lot of it exists mainly in my imagination. I have my excuses, of course: I have books to write, journeys to make. It will never be finished, and if it were I would probably be deeply depressed, but the experience has confirmed a vital principle: houses should be allowed to grow. This street in Poundbury, so superior to thousands of suburban developments, still had the same deadness that comes, I suppose, from having been preconceived and carried out down to the last kerbstone from a planner's blueprint. And that very superiority disturbed me a little too. It had the feel of an upper-class gated community without the gates. But I was being altogether too precious, and when I tried to hint at my uneasiness to the two people unloading their Volvo it was obvious that they really weren't interested in my delicate perceptions. It was only later I learned that the whole subject of Poundbury had been thrashed to death over the years.

'We've only been here a little while, but we're very happy to be here,' she said.

And quite right too. It looked like a very nice house.

'I suppose you have to go into Dorchester for shops and such,' I suggested.

'Oh, no, the village is just round the corner. There's much more to it than this. Maybe that's more like what you're looking for,' they told me, and explained how to get there. So I trundled off along Middle Farm Way and eventually found myself in a village square, full of shops and market stalls and not a single Olde Shoppe that I could see. There was also what I thought was a pretty nice market hall built on plump pillars, which made no pretence of having been there for a few centuries. Well, I liked it, even though I saw later that it had been thoroughly trashed in articles by sensitive architects. Just beyond one corner of the square was a café called the Octagon, where the tea came in pots, so I got one, and a free local paper, and sat down to see who was advertising rooms.

There was nothing in Little Bredy, but there were five or six of them around the area. I compiled a list and phoned them, and they were all full. Some recommended others, but they also were booked up. I started to reach further afield, with the same result, and I began to feel troubled. Then, between calls, my phone rang. It was one of the ladies who'd turned me down earlier.

'I just remembered,' she said. 'There's a young couple who started this year. In West Knighton. It's a bit primitive, of course, but still ... you might be lucky there.' Well, I thought, I'll draw the line at sharing a bed, but otherwise? I phoned, and a woman's voice said, 'This is Sophia.' And I was lucky.

They told me to go up the hill from the middle of the village, and then at the top, on the left, I'd find a red door. Strange how relieved I felt. If I'd been in India, or South America, or Africa, I wouldn't have been worried about finding somewhere to sleep. Something always came along. It always worked out. But here, in my genteel mode, in this land of B&Bs, I felt utterly dependent.

Anyway, it was still morning and Little Bredy was only a few miles away, so I rolled off down the A35, a mid-sized road, then turned off to the left on a smaller road and, before I knew it, I was through Little Bredy and out the other side having passed at the most two buildings, which, even at a stretch, could hardly have accommodated Little Bredy's population of eighty-five. So I stopped and turned around. It struck me that there was not a single person in sight. The countryside was a patchwork of fields, trees and hedges, laid out for me on a lovely summer's day, beautiful and empty. The land fell away gradually to my right in what seemed to be a long, sweeping

133

dale, rising again a mile or so to the south. On my left, just beyond the village sign, a long drive led to a building and I turned into it, thinking I'd better make contact with someone. I'd hardly gone halfway when a car came busily in after me. I had a momentary impression that whoever was in it was not pleased, and the man who got out when it stopped did have an uncomfortably brisk manner.

Then I lifted off my helmet and it changed. Perhaps grey hair has its uses. I explained more or less what I was about and, though he said he was in a hurry, we chatted for a while.

He was far too young to remember anything himself.

'There's a chap up in Winterbourne who was around then,' he said, and gave me a name, 'but probably you should ask Philip down at the house.'

Philip, it transpired, was Sir Philip Williams. The Williams family estate apparently incorporated most of Little Bredy and the house was called Bridehead.

'Just turn right at the junction, you'll see a triangle of greenery, and then go down. The cricket ground is on your right. Just follow it round.'

The 'house' was an eighteenth-century building on two floors with a wide rectangular façade, pleasingly unfussy, of a creamy, almost pink stone, beautifully placed among broad, sweeping lawns. My first impression was of an ambitious and appetising pastry. Its simplicity belied its size: at first sight it seemed square, but I saw later that there was a great deal more house at the back. There was such serenity about the whole scene that I felt quite reluctant to disturb it with my intrusion. I parked my bike in front of the entrance and couldn't help reflecting that it looked rather ridiculous.

I pressed on the bell-push and waited, and waited, and was already moving towards the bike to leave when the door opened. I saw a man I guessed to be in his late fifties, of medium height with greying hair and a pleasant but slightly harassed expression. Nothing about the bike or my appearance seemed to surprise him, but I had evidently caught him at a bad moment.

'We're just leaving,' he explained, as I told him why I was there. It was Saturday, and he was off for the weekend.

'If you want to come back on Monday . . . I don't know what I can tell you. I don't know anything about boys coming here . . . I'll tell you who might remember something, though. That's Arthur . . . He used to be the gamekeeper. He's retired now and lives with his brother, Frank. You might find him there . . . Just go up here and round the stables,' and he described a shortcut to Arthur's house.

'I'm sorry, but as I say, if you're still here on Monday . . .' He shook my hand and went back into the house.

I followed his directions, but with some misgivings. It seemed to me that if this was where we'd been sixty-five years ago he should have heard some story about boys coming to bag potatoes, but I was soon too busy staying upright to worry about it. His shortcut became a rutted cart-track on loose stones, the first time on this bike I had ventured on to dirt. When I emerged, thankfully, at the other end, I was back on the A35, and there were two houses on the corner. They didn't quite correspond to Philip's description, but there were no others, and I spent some time trying to find someone at home. Eventually I ferreted out a woman from the second house who was annoyed at being distracted, and I wondered why these people in this paradise on earth were all so busy.

She knew nothing of elderly retired brothers and was glad to get rid of me. Somewhat dispirited, I decided to go back to CW and deal with a small problem I'd encountered with the charging system, since it was almost on the way to my digs in West Knighton, and everywhere was only twenty minutes from everywhere else. The mechanic who worked on the bike, Richard Downton, had been there at the beginning of my last big journey, and as we talked I found out that he knew all about Little Bredy. Not that he could help me – he was a young man – but the coincidence was somehow encouraging, and in his rich Dorset brogue he told me of an older man who might know a bit more: his father had worked for the Williamses before moving on to another place nearby. As we talked on I had the feeling of slipping back into another age where everybody knew everybody and life still revolved around the big estates, except that now, in the absence of crushing rural poverty, it was voluntary and, in a sense, democratic. So, feeling a bit more cheerful, I thought I'd pick up the trail the next day.

16

The Elusive Potato

Higher Lewell Farmhouse was indeed easy to find. It was at the top of Hardy's Row, where Thomas Hardy's father, I was told, had built some houses for the village poor. It was anything but primitive and, far from having to share the bed, I liked it so much that I was sorry my wife wasn't there to share it with me. Sophia was a lively and pretty young woman who had given up doing something with fashion in the city and come back to do country stuff instead. There was a nice kitchen garden and some happy hens. She was easy to talk with, and talk we did, so it was not long before I discovered that her name was Williams and that she was Sir Philip's niece.

There was a certain genre of fiction dating back to the first half of the twentieth century, such as the mysteries of Agatha Christie or the Bertie Wooster comedies of P. G. Wodehouse, that took place in rural England, where it seemed that everybody was either related to, or acquainted with, everybody else – or at the very least connected in some strange or nefarious way. The girl found drowned in the fish pond was the illegitimate daughter of the chauffeur and the duchess; the sinister man seen at the station wearing an eye patch was the black sheep returned from Australia to blackmail his cousin; the half-witted cleaning lady at the vicarage was the unacknowledged mother of the steel magnate's glamorous assistant; and it was the envious sister of Lady Priscilla who poisoned Lord Arbuthnot's prize-winning pig, and so on. And in the previous century there was no question that the Postlethwaites of Northampton knew how to distinguish between the Entwistles of Chester and those of Bath, because everybody who was anybody knew everyone who was anyone.

Being a city boy brought up among London's millions, I had, naturally, never experienced anything like it and dismissed the phenomenon as an amusing myth or literary device, but now I was brought up against it in reality. Could it be that despite England's rollicking ride into modernity there remained, beneath the surface, a much smaller, cosier rural network of people and families who still lived within miles of each other, and were devoted to their patch? The human equivalent, if you like, of the tracery of little lanes that I was discovering across the country? Certainly big estates like the Williams property, which had survived through centuries, would be a social anchor. Sir Philip presides over 2,500 acres, enough to make Little Bredy his personal village in all but name, but there are other far bigger properties.

I remembered that on my approach to Dorchester from the east on the A31 I had found myself riding alongside a brick wall, which never ended. When I had asked David Wyndham about it he told me it was the Drax estate, and later I read that it is thought to encompass some seven thousand acres, or about ten square miles. The grounds, I read, are opened to the public once or twice a year 'when the local villagers sell tea and cakes', and that's a line that could have come straight out of Bertie Wooster's mouth. According to Wikipedia, the brick wall, one of the longest in England, was built in 1842, and is composed of more than two million bricks. David told me one other odd thing about it: there is a sculpted stag, standing proudly over one of the gates, that has five legs. The story is that, because of the angle of sight, only three of its legs can normally be seen at one time, so the sculptor added a fifth to correct that impression. Alas, Wikipedia reduces the sculptor's genius to banality. The fifth leg is actually a wooden prop to keep the animal in place.

Significantly the Drax wall is only 'one' of the longest. I've been unable to find the others, unless Emperor Hadrian is still a landowner here, but even the Drax estate is small beer compared with, say, the Duke of Bedford's: he has about 25,000 acres, and the Duke of Norfolk 45,000 all told. And even these huge properties are mere trifles alongside the Duchy of Cornwall, which comprises some 200 square miles, including Poundbury and any number of other patches in the south and west of England. I suppose we should think ourselves lucky that the Prince of Wales never thought to build a wall around those.

In terms of size alone, of course, they are dwarfed by land holdings in Africa, Australia and the New World, but those are relatively empty, and their impact would be more economic than social. I remember well my meeting with Mr Gurney back in 1975 on his Nullarbor 'estate' in Southern Australia. It probably rivalled the Duchy of Cornwall in size, but all he had was a small house and a petrol station, since the land was as dry as a bone and vacant but for Aboriginals. But in a country as densely populated as Britain these enormous land holdings must have a profound social impact on rural life.

As I said before, having been brought up in a big city, I had rather ignored the existence of the landed gentry; in fact, I realise that, having had nothing to do with them, I had probably come to some thoughtless assumption that they had dwindled away. There used to be a lot of talk of impoverished dukes having to open their houses to the public or turn them into museums and corporate retreats and such, and I had fallen for it, though I shouldn't have. In 1969, working for the *Observer*, I interviewed the members of a class who had graduated from Eton College twenty years earlier. All of them, bar one, were stockbrokers or gentleman farmers. The exception worked at Tory Central Office. They were all perfectly pleasant people, anxious to put me at my ease and to let me know that Eton had become a quite democratic institution. As evidence of this, one told me that wherever he went on his travels – and he travelled a lot – he was always meeting Old Etonians.

With my own Bolshie background, fed by those stark images of pre-war working-class misery and destitution, I found it difficult to let go of the resentment I felt for upper-class privilege, and I still think it was entirely justified then, but times have changed. I thought I would talk to David Wyndham about it and see how these gentlemen farmers and their big spreads are viewed today. Dave lives about five miles outside Dorchester at Puddletown – it was once called Piddletown but the name was changed to save Queen Victoria's blushes when she came to

visit. He told me that all around his part of Dorset the gentleman farmer is doing very well, and there are numerous big
estates. Not perhaps as well landed as Mr Drax, MP, who also
rides a KTM and is one of Dave's customers, but big enough
to have a village or two.

On the evening I talked to him, he and his wife Rosie had
been celebrating her birthday. It had been a lovely day, and
they'd gone on a five-mile walk, passing through a string of villages all named in part by the River Tarrant, which flows
nearby – Tarrant Hinton, Tarrant Rawston, Tarrant Rushton,
nine of them all told.

'The country is absolutely gorgeous,' he exclaimed. 'All of it
wonderfully cared for, all the fields mown around the hedges,
even the paths mown! You see marvellous old oak trees. If this
were commercial farming, agribusiness, say, there'd be no room
for trees like that, they don't grow fast enough, and the fields
would be ploughed up to the hedges – what hedges were left.'

All this beauty is sustained because the owners can afford to
take a long view, unencumbered by the need to exploit every
square yard of their domain and satisfy shareholders with
quarterly earnings reports. Not that they don't make money,
but it's the beauty itself that brings in the cash.

'Everywhere you see feed containers for the birds. People
come over from Europe for the pheasant, the partridge, the
pigeons and so on,' and, of course, maintaining these great
tracts of land takes experience, devotion and time. It is
impossible to believe that anyone could simply acquire a few
thousand acres of land and convert them into one of these
marvels of sustainable rural beauty, even in a lifetime, even
with unlimited resources. The men who maintain them, like

the old gamekeeper Arthur, work with the skill and experience of generations. Dave told me that the families who own them, who have owned them for ages, can be heard more often these days talking of their responsibilities as stewards of the land.

Should I begrudge them their big houses and the occasional Bentley? For myself, I am more concerned with the fate of our environment than with passing inequities in a society where, for all the huge and growing gap between the rich and the rest, almost everyone is at least surviving. My mother is turning in her grave as I write.

I stayed two nights at Sophia's place, enjoying the village and its very fine pub, and taking vicarious pleasure in her plans to marry a nice, strong, silent man who worked in forestry. On Monday morning I returned to Bridehead. Sir Philip was as good as his word and, when I told him I was staying with his niece, he said his invitation to the wedding had just arrived in the post, and that I should wangle one for myself. He took me into the house and, among other things, showed me a large, airy conservatory at the back where the floor still bore the marks of having once supported a billiard or snooker table.

He did his best to suggest other families in the area who might have benefited from my labours sixty-six years earlier, and printed out a map to guide me. He seemed fairly sure that it wouldn't have been at Bridehead. Then he explained why I hadn't been able to find Arthur. If I had looked across the A35, he said, I would have seen another dirt track leading to a house that was not visible from the road and was half buried among some trees. And that was where a little while later, after a bout of knocking, I encountered Arthur, or was it Frank? To be honest, I got confused. At any rate he was a short but very upright and durable old gent with a bluff and weathered face, in his long-sleeved shirt and his trousers hitched up over his tummy and spilling out over his belt.

I was admiring his garden as he came out, a fine assortment of roots and greens, and we wandered down to the end as we talked. He remembered the depot and told me where it had been, in that same rich dialect. 'Down the road there, in the woods. You'll still see where the huts were, but they took it all away at D Day, see.'

But D Day was in June, before I could have been there.

'And you say you was picking potatoes, but we never lifted them before September,' and that would have been too late for me.

He spoke very deliberately, not pushing it, and I could feel my little fantasy falling apart but . . . I couldn't have *imagined* bagging potatoes, and we couldn't have been there in September because the school holidays surely ended with August, and the depot *must* have been there after D Day because there was my flotation device to prove it. Yet Frank had the utter certainty of a man who had lived his life by the seasons.

So maybe it really was somewhere else, as Philip had suggested, another depot, another farm where they picked their potatoes earlier. All these contradictions and false starts had me developing much sympathy for amateur historians like Alastair Reid. Of course, in my case, it didn't really matter, nothing hanged by it. Even these faltering attempts had given me the window I wanted on this world. I decided to leave it as an enigma.

My general idea was to go west at least as far as Cornwall. A friend had described the Lost Gardens of Heligan, near

Mevagissey, which sounded intriguing. Otherwise I had no particular point to aim at but, for the first time, I began to take in the distances involved. To my surprise it appeared that Penzance was further from London than Edinburgh, and Devon, I noticed, was a really big county, but I had a few days to spare.

I trundled off towards the coast, managing to avoid the big roads. A magnificent landscape unfolded around me, rising and dipping, with occasional glimpses of the sea. A tower appeared on the brow of a hill, and I knew it must be the Hardy Monument, but I didn't appreciate until I got close enough to read the inscription that they would never have raised such an imposing structure for a mere novelist. This Hardy was 'Kiss me' Hardy, the admiral who had knelt beside Nelson on the deck of the *Victory* at Trafalgar, although some say that Nelson actually said, 'Kismet', which seems more likely. There's a profusion of Hardys around Dorchester. An earlier one started a school there in the sixteenth century, but it seems they were all distantly related.

I had read somewhere that this whole area was a favourite place for tourists to come and cycle over, and I am not surprised. It really is gorgeous countryside; the hills are not too severe and allow for constantly changing views. At one point, where the road with its hedgerows seemed almost unbearably beautiful, I thought I really should take a picture, and just as I had stopped to get out my camera, a girl came cycling up the gentle slope. I asked her if she would snap me. It was probably thoughtless of me to stop her on an uphill climb. With a French accent but not much pleasure, she agreed, took two pictures and cycled on up the hill. Annoyed with myself for stopping her, I thought rather stupidly that I might make amends by taking a picture of her, giving no thought at all to how I would get it to her, but imagining that in some way it might become an amusing episode to write about. So, some way after I had passed her I stopped to take her picture, and found myself taking a picture not just of her but of a man cycling behind her who was obviously her father. When he glanced across at me I suddenly saw myself, through his eyes, an elderly man lurking in the bushes, grey hair blowing in the wind, surreptitiously photographing his daughter.

Mercifully they didn't stop. I have no idea how I could have explained what I was doing. This confusion that I have with my own identity is ongoing. The person I see when I look in the mirror is not at all who I think I am, but usually when people talk to me they seem happy to accept me as the person I feel myself to be. Suddenly, and disturbingly, I saw that this might not always be the case. I could get into trouble. Soberly I rode on, digesting this unpleasant information, as I made my way to my next destination, the Smuggler at Lyme Regis, where I had a friend.

17

So Much for Charity

I first met Chrissy Burgess years ago when she and her husband Dan lived in Bridgwater. I was there to give a talk, and they put me up for the night. They had the kind of house that has always appealed to me, with nicely mannered but unselfconscious children. We all got on very well together but I had some kind of special connection with Chrissy, which I was never able to explain. Dan was an enthusiastic sailor so it was no big surprise that they eventually moved to the seaside and acquired the Smuggler at Lyme, halfway up the steeply inclined Broad Street. It certainly had a prestigious look about it. Lyme, after all, is Regis, which is to say, 'Royalty slept here'. Across from the Smuggler, a little way down, is a pub with an ornate clock tower above it, where is inscribed the motto of the Prince of Wales, *Ich Dien*, giving me yet another opportunity to tell my Dylan Thomas anecdote.

When I first visited them there the Smuggler was a restaurant, with rooms above, but what with one thing and another – money, health, exhaustion – Chrissy had let the restaurant go. Now the space behind the big shop windows was full of what looked to me like artful junk.

A friend runs it and apparently it's more profitable than the restaurant and a lot less trouble. In the midst of a depression, for middle-aged couples avoiding the expense of going abroad, the appeal of knick-knacks from a well-stocked junk shop at one of England's prime south-coast resorts could be irresistible. 'I suppose they're all looking for something to take home,' said Chrissy. The pursuit of novelty is key to staying alive in these difficult times, but sometimes it can be over-cooked. I spotted a menu advertising 'Free Range Pâté' but had difficulty imagining the little pâtés gambolling through the glades, as an accompaniment perhaps to *Panorama*'s famous 'Spaghetti Harvest'.

Chrissy and Dan had had one great stroke of luck. They found they could get a permit to build something between their own house and the driveway that curls up to a parking area behind it. So now they had a B&B with parking, a tremendous coup in a place as desirable as Lyme. It was only a small space but Dan, being an architect, had found a way to fit several very nice modern rooms into it, and I was the lucky occupant of one for a couple of nights. Dan was off sailing somewhere, but there were some interesting visitors to drink and chat with.

There was a handsome young couple, Andy and Becky, who were well balanced, intelligent and full of the enjoyment of life, all of which became really interesting when I learned that they spent their working days taking care of the criminally insane, she as a nurse and he as a warden. It was kind of them to spend time dispelling some of my Gothic fantasies.

And then there was a retired teacher from Bath, a solid-looking, bearded fellow called Pat Colbourne, who was busy poking his finger at his fate. Having had a cardiac incident, among other health issues, he had been advised that walking would be good for him, so he thought he'd walk around the UK – but literally, following the coast. His project was called Patsukcoastwalk and it had started at Dover Castle three months earlier in April. Unfortunately one of his knees had turned out to be less solid than the rest of him and by the time he'd got to Lyme at the beginning of June he'd had to take his knee home for a rest, but now he was back, at the beginning of July, and ready to tackle the rest of the five thousand miles ahead of him.

Pat was a biology teacher, and is probably more in tune with his environment than I am but, even so, I think his walk would be something I would find satisfying. From my own walk across Europe I can empathise easily with the pains and the pleasures. I suspect that, unlike me, he knows the names of the trees, shrubs, weeds and flowers he passes on the headlands and beaches. Some time in my youth I rebelled against the academic culture that wanted me to clutter up my memory with thousands of arbitrary names of things, saying, 'I can always look it up.' Now I regret it. Mental laziness has turned parts of my brain to porridge, particularly where flowers are concerned, although I'm OK on vegetables.

Of course, Pat is doing his walk for charity; it's what everyone does, these days. I don't recall people doing anything like that when I was younger. On his website he quotes Sir Roger

Bannister saying, 'Most people don't know what they are capable of,' and I heartily agree, but when Bannister broke the four-minute mile he wasn't doing it for charity. And when I rode around the world on a motorcycle it never occurred to me to do it for charity. I wonder why not.

Perhaps I'm just not very charitable, but something about the whole business makes me uneasy. There's a sweet smell to it. If people want to give away money (and I'm all for that), why does someone have to push an ice-cream cart across the Sahara or bicycle backwards to Burundi to make it all right? From the point of view of the performer, I guess it stiffens the resolve to know that there are people watching. Otherwise, if I were halfway across the Sahara, unobserved, and all my ice-cream had run into the sand I might question the meaning of it all and give up. Doing it for charity, I suppose, gives the goofiness some kind of cachet. You'd never do it if it wasn't for a good cause and, in the end, you get a shot at celebrity. These days, that's as good as gold.

I think another of my prejudices is showing. Why does everything have to be measured by money? It's not that I don't like the stuff, I do, and I'd like lots of it, but I hate working for it. This book that I'm writing now is hard work, you'd be surprised. I don't enjoy doing it, I just love it when it's done. I'll be glad of any money I get for it, but I wouldn't do it for the money. I think it's just to prove that I can. More and more, it seems to be taken for granted that money is an end in itself, but perhaps I've been living in America too long.

People used to do goofy things for the sake of doing them. I'm reminded of the German whom Robert Fulton discovered when he was riding his bike across an otherwise empty Syrian desert in 1932. A twinkling light caught his attention: he veered off to investigate and came across a man holding a mirror in front of him. A sign on his rucksack announced that he was walking around the world backwards. *'Um die Welt ruckwärts.'* As far as we know, Fulton and I, he was never heard of again.

Not that Pat's walk is goofy. Far from it. Walking is a very good thing to do and, given all the ups and downs, walking such a long way at the age of seventy is pretty hard work, so you might as well be doing some good to others at the same time, but then, I wonder, what do you lose? Even though people knew about my journey from the pieces I wrote for the *Sunday Times*, there were long periods during those four years when nobody knew where I was, and that sense of isolation was important to me.

'So far,' Pat said, 'the experience has been wonderful. Beautiful scenery, lovely people, astonishing weather. Time is irrelevant. I just love the solitude of the sea and countryside.'

Being alone is one thing; being lost is another. Thank goodness, we're all different. So, the doing I understand. It's the giving that puzzles me. Travelling around the globe in the seventies, especially in Africa and Asia, I was bound to run up against a lot of charities in the form of NGOs and religious organisations. Some were obviously self-serving; others weren't doing nearly as much good as they thought they were. Often the people working for them were getting much more out of the experience than were the people they were supposed to be helping. A lot of soul-searching has been going on since then, and I believe things were better when I went round again thirty years later, but there's still a fundamental imbalance: money should be accompanied by effort and involvement. Ideally I should take my money to the objects of my charity and help them spend it; then it would be worth ten times its face value. Usually that's impossible, but I must say, Pat has chosen his charities carefully. I don't give away much money and when I do I give it directly to a school I know in Zambia, but if I didn't I'd be happy to have Pat 'Send a Cow' to a farmer in Africa.

Chrissy has Wi-Fi and, with potato picking still on my mind, I thought I'd have another look at the website for my old school. I was fortunate enough (and that seems to be my general condition) to get a really good education at a school in London that was usually called Sloane School, Chelsea, although to tell the truth it was actually in the less prestigious borough of Fulham. We only missed Chelsea by about a hundred yards, so I suppose the fib is forgivable.

I know those hundred yards well. They are on Hortensia Road, which runs north into Fulham Road, and I first ran along them when a V1 buzz-bomb appeared suddenly over the rooftops of Fulham Road and seemed to be aiming straight at me. The second occasion was some years later when a foolish friend allowed me to sit in his car and drive it for the first time. I forgot where the brake was and, to avoid sailing out into the cross traffic, I turned off on to the pavement. The bomb missed me, and the car stalled before it hit the wall. That's how fortunate I am.

Sloane was a grammar school for most of its life and some pretty bright people came out of it before the doors finally closed in 1970. One of them, Mark Foulsham, decided later to honour its memory by creating a website for it, and has done an extraordinarily diligent job. All decked out in armorial

151

colours, it has an exhaustive menu of options, from tracking down old boys (or avoiding them) to peculiar personal ads. There is also an abundance of historical material and Mark is forever delving into the archives. I was rummaging in one of the dusty basements of the website when I struck gold. Mark had news for me:

As soon as war had been declared, Sloane started to play its part in the war effort. Each year, as many as three dozen Sloane boys, accompanied by Mr and Mrs Boas, spent around four weeks on the Bridehead Estate of Sir Robert and Lady Williams, relatives of Mrs Boas (who had become known to the boys as 'Pussy'), at Little Bredy in Dorset. From here they harvested potatoes and other vegetables to help with the war effort. It was also profitable for them and/or the school. The harvest of 1945, from August 17th – September 15th, when the boys were also accompanied by Mr Harry Little, Mrs Green, the school cook, and Mr and Mrs Gurton, they earned £118-6s-6d for a total between them of 2,366 and a half hours! It wasn't all hard work, though, as the boys seemed to enjoy the entertainment they were afforded on the Estate and the extra rations they received for harvesting. They also had to be restrained from accepting gifts from the American servicemen who arrived in the area. The law forbade them from accepting spare military caps, uniforms and weapons, but the school staff realised this was still happening when, one morning, several boys appeared for breakfast dressed and armed as though they had joined the American forces. When the V1 and V2 flying bombs were hitting London in the latter years of the war, Mr Boas arranged for about thirty of the School's younger boys to join him and the others at Little Bredy. They were too small for the heavy work, but helped by potato-picking, blackberrying, weeding, and putting paid to a plague of caterpillars which had been destroying garden crops ... Don Smith recalls that the boys slept in Army tents and ate their meals in the village hall. He saw what they had to do as 'cheap labour' but welcomed the extra money and the use of a full-sized snooker table, which sat in the conservatory at the 'big house'. The tail-coated butler was also on hand to bring them the apples and peaches that were grown on the estate ... It took the War Agricultural Committee four years to recognise the efforts the school had been making, when it offered to provide the tents and equipment the school had already been paying for, along with elaborate printed instructions on how to run the camps.

So there was the elusive proof, and for me there was surprising satisfaction in knowing that I had once scampered across the delicious fields of Bridehead and entered that confectionary castle, under the tolerant eyes of Pussy, no less.

18

Dislocated Memories

The seaside was a big deal when I was a kid. There were songs about it, and you dreamed about it in your dusty classroom, watching chalky motes drifting through sunbeams as you waited for the summer holidays to begin. From London the obvious places to go were Brighton and Hastings. I'd been to Margate and Weymouth as well, but there were other, swankier places I'd only heard about. There was Bournemouth, of course, home of Albert Sandler and the Palm Court Orchestra, who used to play on the radio every week, but that was already behind me. So I decided to aim for Torquay on the English Riviera, where there were real palms.

Going through the list of B&Bs on the Torquay website, I was happy to read that the palms were still there. A frightful row had been going on for some years, with various authorities, like Health and Safety, being accused of wanting to take them all down either because their leaves were sharp enough to cause injury or because people were hiding under them to do unmentionable things, although sharp leaves and unmentionable activities would seem to be a bad fit.

I also realised, when I looked deeper into the map, that the west seemed to be a long way away. Of course, in California where I live you get used to driving long distances – fifty miles for a hamburger is not unheard of – but in England this didn't seem right. If I was going to get to Cornwall in the days I had available, I would have to give up for a while my happy perambulations on country lanes. So I compromised with a red road and charged off down the A3052, only to find that I had been deceived again and it wasn't very far at all.

I arrived in Torquay just after lunch, and found my B&B in a long avenue leading to the waterfront. It had a parking area where a man was washing a car, and I put my MP3 next to him. He looked up at me and said, 'What am I doing?'

I thought perhaps he was offended in some way, but he didn't seem angry, just very alert. He was a small, slight man, middle-aged, perhaps, but with a young face, and he stared up at me with an impish, quizzical look, one of those, 'Come on, try, you can do it,' expressions that make me quite resentful.

'What am I doing?' he repeated.

I don't know about you, but if someone who is obviously washing a car asks me what I think he's doing, I would really rather go somewhere else, because the only reasonable alternative is to stuff his wash rag down his throat, but since I was there to record the foibles of human nature, I said, feeling foolish and annoyed, 'I suppose you are washing a car.'

'Yes, I am washing a car. And is it my car? No, it isn't. So why am I washing it? That's one of the things I do.'

And it came to me that this piece of theatre was his way of telling me that he was not above washing a car for a guest, and I guessed that the play had only just begun. With startling speed he moved on to the next scene.

'I have had six angioplasties in the last two years. SIX!'

I exhumed angioplasty from the mental graveyard where I bury the words I never use, and before I had time to look suitably astonished he was on to the next scene.

'Where was I born?'

This time I refused to bite, and started unpacking my bike.

'Where was I born? Guess.'

I was still trying to put a name to the expression on his face, and at last it came to me. Naughty. He was like a naughty boy, teasing his parents.

'I'll tell you. Hell's Kitchen. Brooklyn,' and then it started to get interesting. He stuck with me, into the house, up to my room, determined to carry some of my things, all the while telling me the story of his life, and always in this strange, rhetorical style, playing the two parts, interlocutor and straight man.

Hell's Kitchen was, maybe still is, an Irish slum in Manhattan. His father, he said, had died when he was three. He himself had scars all over his body from knife and bullet wounds, many of which he contrived to show me. His mother had died when he was ten, and he was shipped off to an orphanage, but there was a relation in England who had agreed to look after him. It had taken eighteen months to do the paperwork. He had arrived, aged eleven, unable to read or write.

156

'How do you think I learned? I had a strict teacher, with a cane. I went from zero to Oscar Wilde in six weeks.'

Inevitably I learned much more about the politics of angioplasty.

'I was lying there, stark naked, shaved from my chest to my knees, with six nurses standing around me, all wearing body armour. And the doctor says, "Ah, Mr McDonald, this is your sixth angioplasty, is it not? I remember you. You were always very nice to the nurses. I have to tell you that these days you don't have to be so polite." And I said, "Quite frankly, the reason I'm polite is because if you fuck up I'm dead."'

Risking the rest of my afternoon, I asked him how he came to be running a B&B. He said: 'You go for a job and— How old are you?'

'Seventy-nine,' I said.

'Christ! My compliments. So – you go for a job and – "How old are you?"'

I was covered in confusion. I hadn't realised he was playing the part of the employer.

'Fifty-five. "Fifty-five? And with a triple bypass? Oh!"'

'We couldn't find any work and all our savings were draining away – so what do you do?'

'Oh dear,' I said.

'Seriously, what do you do?'

'I have no idea what I'd do,' I said, simpering apologetically.

'We bought this place, in desperation.'

It's not what I would have done. I asked, 'You could afford to?'

'No.'

'You got a big mortgage?'

'Two hundred thousand. This place was four hundred and fifty.'

So they'd had a quarter of a million stashed away. Not bad for an orphan from Hell's Kitchen.

And so it went on. But the longer I listened to his dramatisation of his life, and his feelings about his kids, the more I came to appreciate that this was an extraordinarily tough, honourable man.

'Well, we've been here six years,' he said, 'and in six years we've gone from thirty-four businesses in this row . . . ' pause for effect ' . . . to NINETEEN. In six years.'

'But are you keeping your heads above water?'

'Well, we're paying the bills every month. The others have gone bust.'

'Great! Angioplasties must be good for you.'

He laughed, which was unusual. 'Actually I'll tell you something. If you gave me the choice of having an angioplasty or going to the dentist, I'd rather have the angioplasty.'

'It must feel very funny, though?'

'Honest to God, NO. Apart from the prick of the anaesthetic, and the long needles that go into your groin at first, for a couple of minutes, then the bigger one – they tell you to close your eyes and look away because the main needle is as thick as this' – held up a thumb – 'and goes through the groin to the aorta.

'Oh, by the way, there's a couple of dogs. I must warn you. This one's Sam. I'm not sure about his breed, but I think he's mostly Kudalhound.'

'I don't know that one.'

'Well, he likes cuddles.'

I met his wife on the way down, just the kind of nice, motherly lady you would expect – one might almost say cuddly – and I thought, Really, I want them to come through all right. But does he tell his story to every guest? I thought it better not to ask and went off down the avenue to inspect the riviera.

It surprised me. Not by its magnificence. Any real riviera resort would have put on a much more glamorous display of high-rise apartments, hotels, outdoor bars, cafés, restaurants, roving chip and pancake vendors, Andean flute players and so on, but Torquay is low quay (forgive me, I couldn't help it). I loved the openness of it. On my right was a cricket ground, on the left a big park, which in other, meaner resorts might well have been developed. There was no great crush of people, no heavy traffic along the front. I could imagine that it hadn't changed that much since Agatha Christie was still writing her mysteries here. There were a few carnival-type oddities, like a hot-air balloon tethered over a museum, but the general feeling was relaxed and amateurish in the good sense of the word.

I saw a few harmless palms and other trees in a small garden up against a rock face. High above were hotels, and steep flights of steps dangled down the hillside. A notice observed that everything was being redone because it was dangerous. The trees, they said, would be brought back, but then they would say that, wouldn't they? I could see where the unmentionable things might have occurred but those days, thank goodness, are now over.

I had a very nice glass of chardonnay on the terrace of the Pier Point, looked out over the enormous bay and felt very comfortable. The bay was big enough to encompass two other resorts, Brixham at the other end, and Paignton in the middle, so if I got momentarily tired of Torquay I had a choice. To my left was a kiosk offering me another choice, between Agatha Christie and Mackerel Fishing. A wonderful world of opportunity. I went off in search of fish and chips.

Most of my memories fit neatly into the long stream of events, the loves, ambitions, fears, triumphs, disasters, epiphanies, that make up the story of my life as I know it. I remember where things happened, how I got there and how I left them behind. But there are others, a few, quite vivid and detailed, which are strangely sealed off, completely out of context, and no effort can restore the link. One such memory is of a sailing weekend in Africa, on the edge of a large lake. I remember the boathouse clearly, even down to the pattern of the upholstery, but where it was, who took me there and when is an unfathomable mystery. Another time, near my mother's house in Wickford, Essex, I stumbled on an abandoned farmhouse and barn; in the barn there was a huge and most beautifully complicated combine harvester made entirely of wood, with all the other period stuff that had gone with it. But I was a busy man, in those days, and when I tried to find it again, a year or so later, everything had vanished – the implements, the barn, the house, anything to suggest that they had ever existed.

What brought these thoughts to mind was the realisation that I was in Devon. Once, when I had been in the RAF only a short time, I had had a few days' leave. I'd met a pretty girl who was trying to be a model, and she invited me to visit her parents in Devon. Unfortunately for our relationship her sister was also there, and the sister was also good-looking but in a much more extrovert way. Her eyes, I recall, were slightly protuberant but not nearly as prominent as her breasts, and it was immediately apparent to me that I had made the wrong choice. For her part it might just have been sibling rivalry that caused her to thrust herself in my direction, but I was bound by rules of decency that I must have absorbed from books, and I resisted.

It was probably a small house, because there was no room for me, and I had to sleep in the garden in a small tent. At the end of my stay I had a fever of some kind and I was lying out there during the day when the sister came to comfort me in the manner of Jane Russell. But I was an honourable young idiot and resisted temptation to the last. Needless to say, after I went back to Cardington I never saw either girl again. However, that is not my main reason for telling this tale or remembering it.

At some point during that visit I went walking in a nearby forest. It was lovely summer weather, the trees formed a cool, translucent green cavern, and I felt happily at ease. Then I came upon something that thrilled me. There had obviously been a large clearing here, levelled off although it was over-grown, and in the centre was a huge, deep slot in the earth. I am guessing that it was about eight feet across and twenty or more feet deep. I spent hours there, speculating on the kind of industry that would have required this effort. I assumed that it was a mill race, although there was no visible water source. I imagined the enormous wheel that must have been turning there, but driving what? Saw blades? Crushing machinery? Where would the water have come from? How old was it? Who were the people who had populated this landing in the forest? I could almost feel their presence.

When I got back to the house nobody had any idea about it, or of its existence. I had to leave, there was no time to find out more, and later events piled up and buried the memory. Today I have absolutely no notion of where this was. I wonder if the slot is still there. Maybe Health and Safety have filled it in, or put up a sign saying: 'This is a Hole.' What I do know is that I have a great interest in the way things were done a hundred or two hundred years ago, and a sadness at the thought of it all being obliterated.

I compromised with the red roads again, heading for a barbecue with a reader who lived near Saltash, but the next day I was back on the tiny lanes, heading into Cornwall and down to Fowey, which they pronounce 'Foy'. This natural harbour on a rocky coastline was full of excitement when I arrived because the Royal Navy had brought a very modern frigate, HMS *Cornwall*, to visit on National Lifeboat Day and there was a helicopter to rescue people from the briny. Queen Victoria and her prince had landed here too, in 1846, off a very different kind of ship, a wooden one with sails, and it always fascinates me to remember that her ship must also have been seen as very modern.

And then, from Fowey, I rolled down to Mevagissey, which they pronounce 'Mevagizzy', drawn there by the mystery of the Lost Gardens of Heligan. My friends in Stroud, knowing of my interest in gardening, had told me about them. All the way, from one B&B to another, I was plagued with gaudy brochures demanding that I visit the Trouser Button Museum, or the World's Hottest Horse Radish. There was Go-Karting No Speed Limit No Licence; a Treasure Park; a Gem Scoop;

Dairyland – 120 Cows Milked Daily; Waterworld – a
'Splashing Time'; Kidz World; Hollywood Bay; Cyderfarm;
Bearworks; the Tolgus Tin Mill . . . and I resisted them all. But
the Lost Gardens did intrigue me. A family called Tremayne –
and was there ever a more romantic name? – had owned them
for hundreds of years. They had extended over a thousand
acres, and the kitchen gardens alone had been maintained by
twenty-two men, but the First World War had sucked all the
men away and buried most of them, and 'only a few years
later bramble and ivy were already drawing a green veil over
this Sleeping Beauty'.

I am a latecomer to vegetables. In my early days, like most
children, I felt no magnetic attraction. Spinach then was almost
inedible, cabbage was a poisonous green colour, swedes were
just horrible, potatoes were mashed with lumps, and salad
seemed like a waste of time. The idea of raising them was
equally foreign. True, I did grow a suitcase full of radishes at a
prep school I went to for a year, but when I lugged them down
to the stalls on Portobello Road I learned the first lesson of
agriculture: first make sure of your market, then grow.

I didn't dare come back to the subject for forty-five years, but
I had learned my lesson well, and for three years, with a couple
of expert gardeners in the eighties, I sold about half a ton of
organic vegetables a week during the growing season. You
don't grow beetroot with swirls of pink, white and red in them,
or twenty different shapes and colours of pepper and
aubergine, or multi-coloured beans and peculiar roots that go
down two feet, without developing an affinity for the subject.
So the prospect of seeing a huge functioning Victorian garden
fully restored to use was very enticing.

I stalked the beds with notebook and pencil, like a quarter-master, taking an inventory of all the ancient varieties. Since my conversion I have learned how many kinds of food have all but disappeared from our tables because they don't suit the requirements of commerce. A long time ago I read that at one time in England well-to-do families might have a silver platter with thirty or forty depressions in it, each one for a different variety of strawberry; today you'd be lucky to find three. So I was revelling in the roll-call of colourful names. Who would guess that Highland Laddie, British Queen, Rycroft Purple and Snowdrop were potatoes? Or that Shetland Black would be a carrot? And as for Bedfordshire Fillbasket Sprouts, there would be no room now for a name like that in a catalogue.

I was enchanted by the brickwork. What wouldn't I give to have brick walls like those around my garden? And the cold frames, the greenhouses, the hot beds, all wonderful. I retired to my guesthouse, the Mandalay, with a profusion of notes and diagrams, and tips about when to top, and when to weed, and when to mulch. The lady who ran the Mandalay seemed very Cornish to me, though I can't say why. She was very open and inviting, and greeted me with a cup of tea and an accent like Bristol, only more so, but when I tried to take her picture she became bashful and slipped away with her hands over her face. It was a very nice face, and I was sorry not to have it in my camera. I hope nobody was looking for her.

Her place was on the hill above Mevagissey, and next day when I went down to the port, I was very glad of the MP3. The streets around the harbour were very crowded and a car would have been impossible, but I managed to squeeze my little tripod into a space between the marine exhibit and a municipal skip. Despite the crowds Mevagissey was extremely attractive and managed to stay relaxed and enjoyable. There were real boats in the harbour doing real things. I went out along the quayside to the lighthouse pier, where a sailboat was tied up: *Johanna Lucrezia*, two-masted, about ninety feet long and beautifully maintained. The owner's nephew was on deck doing nautical things and seemed happy to chat, so I learned about *Johanna*'s chequered past. She was built by a Dutchman in 1945, supposedly for fishing, but then she was bought by a woman who registered her in England, for cruising, and when that didn't pan out, she was sold on until eventually she became homeless (the boat, I mean, not the woman) and was arrested in 2008 for non-payment of licences and mooring fees to the Inland Waterways Authority. The current owner bought her and had clearly poured his life, soul and wallet into her because she looked beautiful.

At this point a somewhat grizzled elderly chap came by and we walked back together. I asked him if it would be right to call her a schooner.

'Well, she's schoonery, all right,' he said. 'Used to be hundreds of 'em working out of Mevagizzy. Herring and pilchards.' And as one thing led to another I learned the remarkable story of Mevagissey's fishing fleet, and how they sailed regularly to Spain and the Mediterranean with their catch, bringing back salt.

'And coal, too. They weren't your picturesque boats. Black, and in rough seas bloody dangerous, of course. I lost several members of my family.'

They used to follow the herring round the coast, from Falmouth to Plymouth, and on up the east coast to Whitby, bringing in huge catches. Then, in the seventies, it all stopped. No more herring. Fished out.

'Too greedy,' he said. 'Now the French and Spaniards are complaining because they're fishing over there, with seine nets, while the others are using lines. You can't overfish with lines. They're quite right to complain.

'The coal and the salt was all for here, in Mevagissey. People don't realise that boats out of this little port were going all over the world. I've seen documents showing boats going to The Newfoundland in 1601. Took them twelve months, there and back.'

Despite the failure of the herring stock there were still dozens of boats fishing out of Mevagissey, bringing in fish and crustaceans of all sorts, so it wasn't just tourists. Presumably the disastrous floods that hit the port four months later will have been overcome. I can't imagine that in its long history it hasn't learned to deal with a bit of wet. Most of all I am sorry about the herring. What would George Orwell make of it? Well, at least we still have potatoes.

19

A Mutant Village

I had an important meeting in Sussex in a couple of days, so there was no time to go further west. I plucked a couple of B&Bs out of my book, at strategic distances, and started buzzing back through Bodmin, Liskeard and Tavistock, and before I had come to my senses I was in the middle of Dartmoor without much petrol in my tank. I suppose of all the ways of dividing people into two kinds, one of the most common would be between those who are prudent (and boring) and go back, and those like me who are in denial (and stupid) and decide to chance it.

There are not many parts of England where I feel vulnerable, but I suppose Dartmoor might be one of them. For the first thirty miles I can't remember any traffic, or any sign of life. A perfect location for a highwayman to be operating, I thought. Well, at least that scourge has disappeared off the roads of England. Little did I know what would be coming my way.

Somewhere in the middle of the moor I came across a fine hotel, its lounge decorated with *métier* pictures of local farmers and artisans, and they let me know kindly that I was not the only idiot who had charged on to the moor without realising how far he'd come. But they had no petrol, so I ran on adrenalin instead until I came out at the other end on the A30, found a pump just in time, and worked my way round Exeter again and on through Ilminster. The bed I'd found for that night was at a place with the intriguing name of Sandford Orcas, and I'd made such good time that it was only about twenty-five miles away. Wanting to get off the main road again, I took the next turning to the left, hoping to see a village pub somewhere and get some lunch. I found myself going through a village called Martock, and this became a strange, almost surreal experience.

Martock was clearly a village. The houses were small, ranged along the road in the fashion of a village, but as I rode on and on and on, it never ended. I rode for at least a mile still unable to find a pub that was open. Somehow Martock had metastasised. Finally I gave up, turning back to a sort of nursery/DIY/souvenir complex where there was also a restaurant in which I pondered the origins of this peculiar evolutionary twist: the giant village. I must admit the food I got was probably better for me than anything I might have found at a pub, and for the first time I tasted and enjoyed what seemed to be a quite popular drink made from elderflowers, but all this sophistication seemed to belong to a town, not a village.

My computer confirmed that Martock was indeed a village, but with a huge population of about 4,500. Not far away was the town of Somerton, with the same population. What made the difference? Well, it seemed from the maps that, unlike Martock, Somerton had a middle. A town probably has to have a middle. It had also once produced gloves and wire, and it boasted a defunct gypsum mine. Whereas under the heading 'Economy', all Martock could muster was a fish-and-chip shop. Somehow Martock had discovered a way of growing with no central organisation or economy. What is to stop it? If I were Somerton, now only a mere eight miles away, I would be very concerned that Martock might soon come and swallow me up.

Shaking off an eerie feeling, I got back on to the A303 again and, carefully avoiding Yeovil, made my way gratefully to Sandford Orcas, which is a proper village of fewer than two hundred souls and a haunted manor house. It is also very close to Sherborne where there is an abbey I visited countless years ago, and I had a fuzzy recollection of something fine. Next morning I worked my way round a rather confusing one-way system in Sherborne and ended up parking illegally on the pavement because I was damned if I was going to go all the way round again and park legally a quarter of a mile away. Then I entered the abbey and immediately wished I'd left the bike somewhere safe. It took my breath away. Of course, anyone can look it up now, and learn that it's famous for its fan vaulting, but the beauty of it and the lightness, the airiness of that space is incomparable.

I stayed as long as I dared and went on east towards my next night's lodging in Haslemere, south of London. The A30 took me through Shaftesbury and Salisbury – which I already knew quite well – and then I saw that, with a small detour, I could also go through Winchester. Perhaps my experience in Sherborne had sharpened my appetite for architecture, but I knew there was something here that needed to be resolved.

There used to be a song about Winchester Cathedral, which I never really understood, and that may be one of the reasons it has stuck with me through so many decades. Why should Winchester Cathedral 'bring me down'? Well, it seemed absurd that I had never troubled to go there and find out for myself, but here was a golden opportunity. I approached it from the side where there's a Mercure Hotel, and took advantage of its car park, modestly squeezing myself in not to be in anybody's way. The hotel doorman, a nondescript fellow with a moustache, dressed in plum-coloured livery, walked across to me, and I said, 'You don't mind if I park here for a little? I want to have a look at the cathedral.'

'Oh,' he said, with a funny expression, 'that won't take you long, will it?'

What an odd thing to say, I thought briefly, as I walked around the railed-off grounds. At first, being familiar with other cathedrals, I was surprised that there were no soaring spires, but when I entered it I was completely overwhelmed. Twice in one day. All I could do, as I looked down the nave, was wonder how I could have spent so much of my life without ever having been there. The beauty and grandeur of it were just too much for me to deal with. Far from being brought down, I was floating on air. I think, in a way, my feeling was that it was impossible: impossibly long, impossibly glorious, and impossible to describe. I could hear music somewhere at the other end, and found a group of musicians playing something by Ravel, so I sat and melted into the experience. Not being given to superlatives, I find it almost embarrassing to talk about being in an earthly paradise, but that was rather how it felt. Eventually the music stopped and I had to leave.

I did linger with a guide on the way out, reluctant to have to end the visit, and she told me amusing stories about how the whole gigantic enterprise had got going a thousand years ago, and the extraordinary efforts made a hundred years ago to save it, but finally I wrenched myself away and wandered back to the bike in a bemused state. The bewhiskered doorman was still there, with a shorter version of himself, also in plum.

Still in a trance, I said, 'God, isn't it lovely?'

The moustache twitched disapprovingly. 'Don't know. Never been in there,' he said, as though we were talking about a pub or a supermarket. 'Don't care for that sort of thing.'

The other man chipped in: 'Same 'ere. Never been in there either. That kind of thing don't interest me. All I know about it is they dominate the city council. I'm from Lincoln. Don't care for this kind of stuff.'

I could think of nothing to say. The very idea that two men could work three minutes away from one of the marvels of Christendom and never think of having a look was so stupefying that, for a moment, I had to distract myself. What did 'coming from Lincoln' have to do with it? Did he think Lincoln Cathedral was better? It's certainly taller, and brasher but . . . no, Lincoln had nothing to do with it. It was still the same 'kind of stuff'. So what kind of stuff was he talking about? Every now and then I'm forced to remember that there are people I have nothing in common with. It's rare. Like most of us, I have an automatic filter that lets me do business with people of all sorts without ever having to discover how deeply we would disagree about the value of Winchester Cathedral. So I forget that they're there. That's the difference between me and Rupert Murdoch. He never forgets, he always gives them the other kind of stuff, and heaven forbid that he should try to open their minds to anything else.

Obviously there are a lot more of them than there are of me, which is what made Murdoch a billionaire, and me what his media would call an 'élitist'. But what interests me – the reason I'm banging on about this – is whether a society like Britain would even be possible if it were the other way round. Who would stand in a doorway all day in a plum-coloured uniform and do the countless other tedious jobs that our nation has to offer? Or do those tedious jobs exist only to fulfil the needs of the incurious people doing them?

20

Stand and Deliver

A long time ago – so long, in fact, that I might be remembering another lifetime – I became the editor of another magazine and I wanted someone to write about cars. There was very little money about, so I was lucky to find someone who didn't need any. Sir John Whitmore had inherited his baronetcy from an ancient family, along with a lot of money and land, and a large pile of masonry called Orsett Hall in Essex. His passions in those days were driving fast cars even faster, and spending the nights in discothèques. He became British saloon car champion in 1961, drove Le Mans with Jim Clark, and generally tore up the town.

However, John was not quite the 'twit' (an affectionate nickname) that he may have seemed to some. He had strong opinions about cars and driving, and was looking for an outlet. Since my magazine was in the *Playboy* style, it was a natural for

John, and one day he found his way to me in my office in Salisbury Square. I met a tall (very tall for a driver), lean man with a pleasant but assured manner, a narrow face with blue gimlet eyes, and a beak-like nose to match his rather nasal upper-class voice. Together we fashioned a very useful monthly column, and later he brought Jackie Stewart in as well to write a racing column for me.

I abandoned the magazine in 1967, but we stayed in touch, and in the following years both of us changed our lives quite dramatically. John discovered the New Age and explored it relentlessly, from Findhorn to Esalen, financing films and all manner of exotic experiments in mind/body research and extra-terrestrial communication. Eventually, when he had pretty much disposed of his wealth (which many thought was one of his objectives), he found a profession in coaching, which came out of his interest in Inner Game Theory and now, forty years after we had first met, he had a world-wide reputation as a business coach. This was a man who would surely have an interesting perspective on how England was faring these days.

Between his many trips to other continents it was no easy thing to find him at home, and that was why I had come back from the West Country more quickly than I would have preferred. We were due to meet at his place in the early afternoon and, because he lived not far from Tunbridge Wells, I thought I would have a look at this famous spa town. At first my TomTom would not co-operate. It refused to admit that the town existed until I remembered that the correct name was Royal Tunbridge Wells, when it condescended to usher me in.

The area where various royal families have disported them-selves most often is known as the Pantiles, and it does make a lovely scene of Regency buildings, with a terraced colonnade, all painted a gleaming white. Unfortunately there were several thousand other people enjoying it at the same time, and park-ing, as usual, was a pain just about everywhere I went. Finally I found a small side-street with a coffee shop at the corner, and seeing a motorcycle parked on the pavement with a chain on it, I put my MP3 alongside it and went for a coffee while I checked my emails.

When I came out, after about ten minutes, and turned the corner I saw that something yellow was attached to one of the front wheels. I had never been clamped before so it took a moment to sink in, and then I became quite annoyed, assum-ing naturally that some traffic warden was being over-zealous. I walked across and started to look at the notice attached to the bike, scarcely able to comprehend what I was reading, since it read like a satire from *Private Eye*.

It said: 'The vehicle has been wheel-clamped in accordance with conditions displayed and is subject to a charge of £90 per calendar day/part calendar day parked, and a release fee of £140, all payable prior to removal of the device.' It said I had been photographed and, among other things, that 'time-wasting delays in making payment will be charged at £35 per part quarter-hour. Further "call-outs" will be charged at £80. Once a tow truck has been called, tow away fees will apply. Thank you.'

I was in the middle of reading this farrago of nonsense when a man came up from the other end of this short street and, in a deliberately calm voice, began to spell out my predicament.

It seemed that this was not, after all, a practical joke. He was a young man in good physical shape, and although he wasn't in uniform he had the look and the bearing of someone who normally would be. He wore black trousers with an ankle strap, and black boots, and a long-sleeved grey shirt. His sandy hair was short and showed the remnants of a fading Mohican cut. His eyes I remember as being grey and very cold.

He told me that I was parked on private property and that the various charges I would have to pay, if I didn't want to be towed away, added up to £390. This was so absurd that I couldn't take it in. There was nothing about the street to distinguish it from any other, nothing at all to tell me it was private, until he pointed to a notice on the wall opposite that certainly didn't attract attention. There was a charge for parking, he said, and a charge for being clamped, a charge for calling the tow truck, and if I waited, there would be another charge for being towed. If I paid him £390 now he would be kind enough to send the tow truck back.

It seemed inconceivable that this could be happening. If a highwayman had leaped out of an alley with a pistol and demanded £390 I would have grasped the situation in an instant. But this seemed so far beyond the bounds of any expectation as to seem insane. How could anyone simply seize a piece of normal city road and say, 'This is mine, pay up'? I walked over to read the notice and it confirmed everything the man had said, and yet, I thought, I could just as easily put a notice like this up anywhere and fleece anyone who came along.

'If you don't have the cash,' he said, 'there's a bank machine round the corner at the other end.'

Very considerate.

I sat on the bike and phoned John.

He was as stunned as I was. 'This is not right,' he said, several times.

'I know, but what can I do? I can't just sit here and wait for a tow truck. There's nobody around. I thought of calling the cops, but he says they've been around here before, for someone else, and they can't do anything because I'm on private property. And I can't do anything with this chunk of steel chained to my wheel.'

We tried to think of something to say but it all came to nothing.

'Well, we won't let this go,' he said. 'We'll definitely follow this up. I've got good connections with the police, and with the council. It can't be right.'

'OK,' I said. 'I'll see you in a bit.'

All the while my mind was spinning frantically, trying to think of some way to crack this thing open. It was extraordinarily humiliating to be rendered so completely helpless. Although no one had laid a finger on me, I felt physically violated. If I'd been a lot younger and stronger, I might even have tried violence myself, but something about this man suggested he would be ready for that.

Was it the money? What if he had asked for five pounds, or even twenty-five pounds? I would have paid up, with a wry grin or an angry remark, and put it down to the cost of doing business. But £390 was beyond all reason. I couldn't stomach it. I was choking on it. And yet I've lost much more than that in the past and taken it in my stride. No, it wasn't the money. It was the bare-faced injustice of it. It felt like highway robbery, with the robber standing right in front of me, and I couldn't touch or stop him.

I walked to the other end of the short road. There were no houses on it. On one side, the side I was parked, was the high brick wall of a multi-storey car park. On the other side was some sort of industrial building and some steps leading up to a door. Sitting on the steps, smoking, was another man in very similar clothing, also young and fit, and looking pleased with life. I got cash from the machine and came back to pay the man.

'Do you get a commission?' I asked.

'Yes,' he said.

'Quite a lot?'

'Yeah, well, it's not nice work. Somebody has to do it.'

Then he took off the clamp, and as he loped away I took a picture of him, just to prove that the whole thing wasn't a figment of my imagination. I couldn't let go. I rode back to the Pantiles, where there was a tourist information centre, to see what they had to say. A woman there knew nothing about it, and was cautiously surprised, in case I'd made the whole thing up. They had a map of the city and I found the spot but the fateful road wasn't even on the map. It could have been the road to Hogwarts.

John's home was strangely difficult to find because of a confusion of names and I had to drive around in circles for a while, so that by the time I saw him waving at me from the corner of the street a scab had formed over my psychic wounds. John was eager to get to grips with the Tunbridge Rip-off, so he got me to write out a statement and we spent some time photocopying it and the bits of paper I had got.

John had been writing a motoring column for the *Daily Telegraph*, and being Sir John made it easy, of course, for him to know all the relevant notabilities, like police commissioners and councilmen. It wasn't long before we discovered that this man-sized fly trap was well known to them, that others had been caught in it, and indeed one poor soul had lost his car in it because he hadn't been able to come up with the money before it had started mounting up.

What was more, this flytrap wasn't the only one of its kind in the country, apparently. John was determined to pursue it, and during the rest of the time that I was rolling through the British Isles, he gave me frequent updates on progress. The police considered themselves helpless, and I found this difficult to swallow. Since it would be very hard for anyone to guess that that particular bit of road was any different from any other, and the amount of money demanded for parking there so disproportionately large, I simply couldn't believe that these people weren't guilty of some kind of extortion or misrepresentation.

The council, on the other hand, had some promising news. The multi-storey car park behind the high brick wall where the foul deed had been done was council property. Prompted by the fuss that John and I had kicked up, they looked into their archives and they were now convinced that the road itself also belonged to them. They decided to take the matter to court; if they won, it was said, I might get my money back. I decided not to hold my breath, but John asked them, on my behalf, to put up some very visible signs to warn others of the danger they were in.

I tried to compose some words to commemorate my historic

tussle with the pirates of Tunbridge Wells: 'Here, on August the 15th, in the Year of our Lord 2010, Ted Simon Succumbed to an Act of Foul Banditry . . . etc, etc.' After all, these large enamelled notices, spaced along the wall, are as close as I shall ever come to having plaques raised in my memory, but the council's signs were up before I could get the wording right.

21

How I Got into the Papers

I had thought, at one time, that I would write my story around
the A6, the road I identified with most strongly during my
hitch-hiking days, but I see now that I am mistaken about the
route my drivers followed. Although they started on the A6,
and my journeys to Glasgow also ended on it, to have changed
lorries in Warrington I must have been taken on a wide diversion, probably through Macclesfield. The road signs were
pre-war and vestigial and nothing had been done to improve
them – partly to confuse the Germans. I would have had to ask
where we were, and no doubt I was asleep some of the time. In
any event, although the A6 goes through Manchester, I didn't.

So it seems that I had never actually been to Manchester
until now, and yet Manchester had always been a presence, the
most dominant presence of all the English cities, because of
the Industrial Revolution, because of the paintings of L. S.

Lowry, because of the Manchester Ship Canal and the Hallé Orchestra, because, as I mentioned earlier, of the malign reputation of the news editor of the *Manchester Evening News*, and because of the relatively inappropriate civility of the *Manchester Guardian* and the elegant prose of Neville Cardus, among others.

All of this created an image in my mind of a great city bursting with energy and big, bustling crowds, so I was completely at a loss to find that Manchester was much smaller than I had imagined, that half of it turned out to be Salford anyway and, most puzzling of all, it seemed strangely empty. Of course that, in its own way, was perhaps a blessing. I could walk around with a sense of personal freedom that is sometimes lacking in London. I was very fortunate to have as my only friend in Manchester Dan Walsh, a man who made all other friends redundant. Not only does he know almost everything about the city, but he is poetic with it. He lives in Hulme, a quiet spot with a park-like feel yet very close to the centre, and there was a bed for me too.

We walked around a lot, looking at old industrial buildings that had been converted into offices and flats, some very prettily. All of them had roots in this country's extraordinary history of invention and development. One of them was where Thomas Hancock manufactured the rubberised material to make Mr Macintosh's famous raincoats. In fact, but for a quirk of Fate, we might all be wearing Hancocks, and Apple computers might not sound so appealing. When I said Dan knew almost everything, I was thinking of some strange cast-iron hooks that protruded from the outside walls of that old factory. They stuck out several feet, in rows, and baffled us both, but he had now plugged this hole in his omniscience. They once carried pipes that sucked out the noxious vapours from the rubberising process inside, presumably to poison passers-by instead. Second-hand rubber.

We went to the Lowry Museum, of course, and I gazed over the Ship Canal that in its day, I suppose, was a grimy basin of sweat, hope and misery but is now a gleaming arc of water through the city. In today's money it cost more than £1.25 billion to build, which gives one an idea of the stupendous wealth that was generated by Manchester's textile mills. We mingled with the crowds on the only busy street I remember seeing. We visited markets and restaurants, and pubs that told stories, and I thought, There's nothing wrong with this. I could live here.

On Saturday morning, rather reluctantly, I wheeled my MP3 out on to the road north, and followed it through Salford, past the cathedral and the rows of cottages, and on to the A6 past Preston and almost to Kendal before turning left for Barrow-in-Furness and the next encounter with my distant past.

I stayed on the main road as far as Ulverston, but then I cut south to follow the coast along the edge of Morecambe Bay, coming into Barrow from the back, and found myself very quickly in the heart of town. There was that opulent Victorian town hall in dark red sandstone, and the broad main street, still vaguely familiar after fifty-six years. By some mysterious method of calculation I can't fathom, I had decided I deserved to sleep in luxury that night, and I had chosen a room at the Abbey Hotel. The people who advertised it on the web called it 'historic'. They didn't say why, though there was apparently a 'romantic' abbey alongside. I found the hotel eventually, a long way down, on the continuation of that same big main street, an imposing three-storey building in crusty brown stone, about a hundred feet long with vertical features dividing it into rectangular sections, which gave it the look of fortifications, all set in lovely grounds. I went to my room and tried to devise a plan of campaign.

Somewhere around 1953 Britain raised the age limit for National Service from twenty-six to thirty-six. I wasn't sure why; probably because of the Korean War, where we had some 60,000 troops. What it meant for me, hiding out in Paris at the age of twenty-two and having a hard time surviving, was that it wasn't worth the candle. So I returned to England and offered myself to Her Majesty. To my surprise, the Queen did not jump at this golden opportunity. On the contrary, through her representatives she let me know that I could jolly well wait until she was ready, which was a nuisance (no doubt well deserved). After all, who wants to employ someone who might be snatched away at any moment?

Pretending that I was an ace journalist with Continental experience (which was about three feet away from the truth, since all I knew about journalism I had learned looking over the shoulders of real journalists), I applied for jobs on prestigious rags such as the *Baker's Gazette* and the *Kensington Journal* but they either saw through me or past me. I was lucky, in the end, to get a job on the *North-Western Evening Mail*, a paper published in a place I had never heard of called Barrow-in-Furness. Going there tested my mettle much more even than going to Paris because, like most Londoners, I thought of the 'north' as being little better than Siberia. They needed a sub-editor and, because it was an awful long way to Barrow, they dispensed with the usual interview and said they would simply get rid of me after a month if I turned out to be unsuitable.

Sub-editing is a skill made in heaven. If you are born with it, you can learn to do it in an afternoon. I proved this by learning to do it myself before they found out I couldn't, and later by teaching a friend with no newspaper experience at all to do it in a few hours. Overnight he became a sub-editor on a national daily paper and thrived there for years, if you can call life on the smoke-laden, booze-sodden news desks of those days thriving. And actually I can.

My own entry into this disgraceful profession was even more colourful, and the result of pure chance. Before deciding I would escape abroad and become a great novelist, I played jazz clarinet, badly, and haunted a club in Oxford Street where the Humphrey Lyttelton band performed. His clarinet was Wally Fawkes, a very warm and generous man, who also drew a famous cartoon strip in the *Daily Mail* called Trog. He deigned to speak to me, and when I told him I was off to Paris, he gave me a note to a chum who managed an offshoot of the paper called the *Continental Daily Mail*, and this equally nice man gave me a job as a messenger boy.

My duties consisted of standing in a corner of the office next to a short, forty-year-old Frenchman called André, with liquor on his breath and a Gauloise stuck to his lip. In front of me, in a wheelchair, sat Sandy Dougal, a war invalid, who was the copy taster. This is – or should be – a job of grave responsibility, since the copy taster separates the wheat from the chaff, and has to be counted on not to throw out the news that the president has just been shot or, more importantly, that the tax on alcohol has been suspended for a day.

Mr Dougal and I were the first to arrive in the office, in the early afternoon. He tackled the rising mound of dispatches telexed from London and I fetched cigarettes and sandwiches, but one day he didn't appear. When the dispatches began to overflow on to the floor I decided, having looked over his shoulder for a couple of months, that I would do his job for him. This seminal stroke of *chutzpah* changed my life. Dougal never returned, and they let me go on doing his job. Unfortunately they shut the paper down three months later, but that was how I got into the newspaper business.

Of course, this is very different from becoming a journalist. That takes many years and is not something, in my opinion, that can be learned at school either. I'm not sure that I ever mastered it myself, despite the exalted position I arrived at. I'm too partial, and a bit scared of putting my foot in the door.

Anyway, back to Barrow. When I got off the train in the afternoon, the sun was shining and it was quite warm, which confounded my prejudices and cheered me up. I remember actually singing on my way to the newspaper office – 'Once I Had A Secret Love' and 'Three Coins In A Fountain', because I had a crush on Doris Day at the time. The office where the subs worked was quite cosy, just large enough for five or six men sitting at tables. The only things that distinguished it from any other office of the day were the spikes, six inches of wire with a sharp point sticking up from a wooden base, on which we disposed of surplus copy. One became adept at slamming paper down on this thing, but I never heard of anyone impaling himself, except in crime fiction. 'Spike it' became a colloquialism for disposing of something, but I don't know if it has survived the spike itself; I haven't heard it used myself in a long time.

An MP3 is just what you need to roll around a congested but delightful port like Mevagissey. It has real boats doing real things. It also has the *Johanna Lucrezia* (below); very beautiful, very 'schoonery'.

There was this red bike parked on the pavement, so I put my MP3 alongside it and went round the corner for a cup of tea. Next thing I knew the clamper from hell was walking off with £390 of my money, in cash. I fought valiantly, to no avail.

PLEASE BE AWARE

THAT PERSONS PARKING HERE
WILL HAVE THEIR
VEHICLES CLAMPED BY A
PRIVATE CLAMPING COMPANY.
TUNBRIDGE WELLS BOROUGH COUNCIL
IS AT PRESENT IN DISPUTE
AS TO THE RIGHTFUL OWNERSHIP
OF THIS LAY-BY

Finally, the council of Royal Tunbridge Wells raised these plaques in my honour.

I have several times been a guest on the barge that Lois Pryce and Austin Vince keep moored over West Drayton way, but it was a casual glimpse of the Grand Union Canal (below) as I was riding through the middle of London that finally woke me up to the water-ways of England – another world floating peacefully and secretively among us.

The two faces of Barrow-in-Furness: the pomp and magnificence of what Barrow might have become and, hardly a stone's throw away, the working class town that Barrow is today. Not that there's anything wrong with fast food from Istanbul, or a wonky heart.

I used to hitch through Carlisle but never got a chance to stop and look at this nice university town, pleasant despite the awful warnings.

St John's Tower in Stranraer, a lovely capsule of history (below, right).

ANTHRAX
IS KILLING HEROIN USERS
ACROSS SCOTLAND

EARLY TREATMENT CAN SAVE YOUR LIFE.

If you suspect you may have anthrax go to your nearest A&E

If you want advice on local drug treatment services contact:

Know The Score

0800 587 587 9

www.scottishdrugservices.com

Drew Millar knew full well that I could never leave Northern Ireland without a pilgrimage to Joey Dunlop's bar, even though I had only a vague notion of his accomplishments. But that was soon rectified.

By the banks of the Liffey – the water that turns to Guinness in a jiffy (below).

Billy Scott and his magic red cab gave me the ultimate Belfast experience (above left).
The wonderful wind-blown statue of one of Ireland's heroes,
John Barry from Wexford (above right).

The ultimate accolade: a picture with Joey Dunlop's widow Linda (below).

We're all enjoying an Ulster Fry at St George's Market. No healthy options for me, thank you very much.

An unexpected windfall: the Vauxhall police car, complete with coppers, that was the scourge of the lorry drivers in the forties. Pictured beside the Lagan, in Belfast (below).

The editor had a small office, and that, I think, was that, apart from the classifieds, which were really the heart of the paper. The most important news feature was a daily report on the fat-stock prices. As long as you got that right, nothing else mattered very much.

The other subs in the room were older, of course, and had come back from the war. The only one I remember had been a special-forces major, who had parachuted into the mountains of Yugoslavia to fight with the partisans. He didn't speak Serbo-Croat, naturally – there can't have been more than a handful of people in England who did – but he said quite seriously that as long as you shouted loudly enough at them in English they understood.

I found lodgings with a married couple in a long street of identical semi-detached houses. The husband was a small, balding, inconspicuous man, who worked all day and sat in an armchair through the evening. She stayed at home. She wouldn't have been bad-looking but for her false teeth, which were a common feature in those days. Teeth in Britain had been generally bad for generations, and many women were advised by their dentists to have them all out once they'd got their man because they were more trouble than they were worth. I didn't understand how dissatisfied she was with life until one morning, when I was lying in and he had gone to work, she came into my bedroom and started to tell me how excruciatingly bored she was with her husband, who 'just sits there and never does anything'.

She sat on the end of the bed and somehow contrived to arrange herself so that stocking tops, suspenders and white flannel knickers were all displayed as though by accident.

When I think about it, it would probably have been good for me to have it off with her, but the thought of it (if I even gave it a thought) would have made me ill. My aesthetics have too often got in the way of sex (and sometimes led to disastrous sex too). Anyway, in the five months I lived and worked in Barrow, I can't remember having sex with anyone except myself. A tragic waste, really.

I left the frustrated housewife after a few weeks to share a flat with the paper's solitary reporter, Bob, a man of about my age, and through him I found out a lot more about the town and the country around. The history of Barrow – or, rather, his version of it – was interesting and poignant. Back in the nineteenth century it was a natural jumping-off point for Ireland (did I mention that it's on the Irish Sea?). It flourished on trade and the city fathers fancied themselves as rivals to Liverpool. They built a wide avenue and imposing Gothic buildings in the centre and confidently waited for the wealth to spread, but they didn't count on the cupidity of the Furness railway family.

Key to the prosperity of Barrow was the railway, which connected it to the great national railway system, and brought the goods to and from the port. The Furness railway was a relatively small enterprise and one of the big lines wanted to buy it, but the owners, who thought they had a stranglehold, held out for a price that the big guys thought was ridiculous: they gave Furness the finger and built a completely new port down the coast called Heysham. Which literally stopped Barrow-in-Furness in its tracks.

Well, it's a good story, but it doesn't take account of the fact that at one time, towards the end of the nineteenth century, Barrow had the largest steel works *in the world*, while alongside it a huge shipyard grew and grew through various incarnations. When I was there in 1954 it was called Vickers, and I remember visiting the port when some destroyers were being delivered to the Venezuelan Navy, but what I know about the company I have learned since. Many famous ships and submarines were built there, and the interests of Vickers spread and embraced just about every warlike icon of heavy industry in Britain, from the Maxim gun to the Spitfire. Eventually, and during the years of my absence, Vickers turned into BAE Systems, with an even more colossal establishment, but in a sense you could see the truth of my flatmate's version, because Barrow was totally dominated by steel and ships. The free flow of commerce that brings variety and sophistication to a great port like Liverpool was missing. Barrow was, and is, a company town. Beyond that small, pompous centre there were just rows upon rows of workers' houses and tenements. Pubs and working-men's clubs were the entertainment.

Even so, though life was drab it wasn't that bad in 1954. Australia sent a shipment of food to relieve Britain's depressed areas, and Barrow was on their list. I went out one Sunday to a council estate with a truckload of these victuals to watch an ambitious councilman distribute the largesse. He was naturally hoping to get some second-hand benefit from the gratitude of his parishioners. Grabbing the biggest tin he could see, he went to a ground-floor door, knocked and waited, with a big smile on his face. As we passed the window I saw a family seated inside already eating a large roast dinner, and I lingered to watch the wife rise to answer the door. When she opened it he thrust the tin at her, grinning, and said, 'With the compliments of the people of New South Wales.' She looked at the tin, nonplussed. He hadn't bothered to notice that all it contained was potatoes. She did not need potatoes.

Although life was more fun in the flat I still never managed to get a girl. Perhaps I was ruined by three years in Paris. My flatmate, on the other hand, did well for himself. It has struck me since that men have an amazingly large vocabulary of demeaning words to describe girls. There's 'skirt', and 'bint', and 'crumpet', and 'pussy', and 'doll', and 'baby', but Barrow is the only place I know where they were called 'boot', and I didn't have the sense to ask why. Bob had his boot, and I had to spend evenings out while he entertained her, so I know just how thin the public offerings of Barrow were in those days. The absolute zenith of culinary luxury was an expensive mixed grill at the caff, and if you could distinguish between the ingredients you were lucky. Wine? You must be kidding!

Bob's reporting took him all around that part of Lancashire and Cumberland, and much of it was anchored in the distant past. He told me that it was still not uncommon for parents to sew their children into their underwear for the winter, and I am indebted to him for one of the most touching stories I know. In a hospital in Whitehaven a nurse told him of an elderly woman in her care who wore a nightdress she absolutely refused ever to take off in the presence of others. On the front of the gown was an area of beautiful floral embroidery with a hole in the middle of it. She explained, 'It's where me 'usband cooms t' me.'

Thanks to Wi-Fi, I was able to discover that my hotel, Abbey House, had been designed, a hundred years earlier, by Edwin Lutyens, who had also done the Imperial buildings of the Raj in New Delhi, among other famous edifices; in other words, a great architect. It had been built to house a flat for the managing director of Vickers, miles away from the sheds and factories he oversaw, and as a guesthouse for clients and visitors. In effect it was a castle for an industrial barony, a measure of its power and significance. I had certainly never seen it in 1954, and it was the first of many indications that I had never really seen anything much at all. I explain it in part by my lack of transport, but mostly by an unpardonable lack of curiosity. What I saw on my daily walk to and from the newspaper office gave me no inkling of what lay beyond on the dock side of town.

By now I knew that I had shot myself in the foot by booking the ferry to Ireland on Tuesday because it meant that I would have to leave Barrow on Monday morning, and there would be no chance of revisiting the paper's offices. Obviously they would be closed on Sunday when they didn't publish, and it was already too late in the day to expect them to be working. I tried the phone and got a man who sounded young, dull, unhelpful and quite indifferent to my story, so I gave up on it. After all, there was little to be gained. There could be nobody there I knew, and these days there's not much to distinguish a newspaper office from an insurance company – there's nothing to see but computers. Instead I spent the afternoon rolling around the port area on my MP3, and across the bridge to Barrow Island, increasingly amazed at what I saw, and ashamed for what I had missed all those years ago.

Most of my journey so far had been preoccupied with old stuff, ancient ports and monuments, cathedrals, abbeys and the restored carcasses of the Industrial Revolution, so it was even more shocking to have come to this backwater, this cul-de-sac, which some even called the armpit of Britain, to find myself confronted by a huge array of the most modern industrial power imaginable. The old Vickers, transmogrified into BAE, had spawned a dozen industries housed in enormous gleaming hangars all devoted to the one end that trumps even the worst recession: warfare. The biggest of all the buildings, the Devonshire Dock Hall, is a modern version of the Cardington hangar, with those extra feet added to avoid having to clip *Titanic*'s toenails. I think it's the biggest thing this side of Dubai. Battleships, nuclear submarines, anything enormous can be built inside it. All around it other vast, pale sheds proclaim that they serve 'Global Combat Weapons Systems'. Better to be killed by a system, no doubt. Nothing personal.

Guns, bombs, drones, torpedoes, all smarter than hell, are built here. And then, when you have driven past this dazzling display of Britain's most profitable, cutting-edge export industry, you find yourself among the houses that were put up more than a hundred years ago for the men who swarmed into the Barrow shipyards at the beginning of the age of steam. And the men who build all those marvels of murderous technology apparently still live in them. Since 1845 they have been sending hundreds of warships, submarines, tankers and liners out into the oceans, including, as it happens, the SS *Oriana*, which took me to Australia in 1975. There are two nuclear-powered Trident submarines under construction now, by an offshoot of BAE called Submarine Solutions, but the houses remain, superficially at least, as they were.

Not that that means they're bad. I found myself between

long unbroken terraces of brown stone houses facing each other. All the stonework was neat and dark, and even on this bright, sunny day the houses seemed to be waiting impatiently for winter. No doubt when they were built as workmen's dwellings they would have been exactly what was missing in Wigan and the rest of industrial Britain. They must have been quite advanced for their time, but there was an indefinable feeling of claustrophobia, like being trapped in a maze.

They reminded me of a much more depressing project I was lucky enough to see in London before it was torn down, and that was Columbia Dwellings in Shoreditch. It was built in one big U-shaped block by the philanthropist and banking heiress Angela Burdett-Coutts. Prompted by Charles Dickens, she had intended to uplift the spirits and morals of the poor. Of course, it must have been vastly preferable to the gin rows it replaced, but when I saw it a hundred years later it had a black, miserable and preachy aspect with its towering three-storey Gothic arch recessed into the brickwork over a tiny entrance where the human ants would come in and out. The inhabitants would have none of it. It was completely vandalised and urine-besmirched. These houses in Barrow, however, were apparently well kept, perhaps because they were houses and not tenements, and also because they were associated with work and completely outside the mainstream of society.

Two men were seated next to each other on one of the low walls. One, in trainers and a loose green football jacket, seemed oddly disconnected, looking round vaguely. The other was much younger and the thought came unbidden, and probably wrong, that they might be dealing. I stopped and asked if they lived here.

'Well, kinda,' the older one said. 'I've just got back from Thailand m'self.' Here in Vickerstown the very existence of Thailand seemed improbable, just as Venezuela had seemed so exotic years ago.

'Holiday?' I asked.

'No, work.' Laconic.

'So what's it like to live here?'

'It's hard to get out,' said the younger one. 'There's work here, you see, but out there ... ?'

The older man stood up, waved, 'See ya,' and walked off. The younger man followed him. I rode back to the town hall, parked behind it and went into the shopping areas on the other side of Duke Street.

There were public works going on. The council was following the fashion, digging up the streets, laying paving stones and turning it all into a pedestrian zone, but the shops were small and sorry-looking, and several had given up. Others were the kind that spring up in a depression, like second-hand shops with hand-painted signs. An Asian man with a travel-goods shop was pulling down his shutters early.

'Business is slow, of course,' he said, 'but we are hoping soon that it will pick up.'

He didn't look very hopeful. I asked whether he thought making it a pedestrian precinct would help.

'Perhaps in the long run, but I think maybe they are jumping on the gun.'

The area around the original town centre is actually another planned town, this time a Victorian one. The tens of thousands who flooded into Barrow in a couple of decades to man the immense ironworks and the shipyards obviously created an instant demand for housing. A man called James Ramsden, who first ran the railways and then almost everything else, was up to the challenge, and designed the heart of what he probably imagined would become a metropolis. He was heralded as a local hero, made mayor five times and was knighted for his efforts. A fine statue to him stands in the middle of a rather attractive square, which he must have had a hand in planning. One wonders whose statue he thought might grace it. He needn't have worried: his own went up twenty-four years before his death. At the next big crossroads there is another square, complementary to the first, where there's a statue to the other hero, Henry Schneider, who started the railway, discovered new iron deposits and brought the steel to Barrow.

There was little to do in Barrow in 1954 and not much more now. The council has built a nice auditorium called the Forum, but *Peppa Pig's Party* on the silver screen was not my cup of tea, and I had chosen the wrong day for me to Sing My Heart Out and Boost my Immune System. The café where I once had my mixed grills was gone beyond recall so, in the absence of a boot to cheer up my evening, I settled for a night at the Abbey Hotel, where I wandered around the building and tried to reconnect with the ghosts of Vickers past as they glided past me over the oak-parquet dance-floor. Then I made a complete pig of myself on excellent food and wine and complained inexcusably over what were really quite petty mistakes before hauling my sorry carcass off to bed.

22

Where Did the Taste Go?

The presentiments of rain that had been hanging over Cumbria during the last days were gone by morning, and I set off in bright sunshine to circle Lake Windermere on my way back to the A6 and to pay homage to the Jungle Café.

The Jungle Café is one of my fondest memories and yet, in truth, I can remember almost nothing about it beyond the huge sense of relief, warmth and comfort I felt when we went in, and the sheer pleasure of being given a full, substantial breakfast with nothing spared, nothing rationed. There were real eggs with deep yellow yolks, thick meaty bacon, bangers and mash, mushrooms, tomatoes, baked beans, fried bread and gallons of hot, sweet tea. If I'd had no other reason to sit in a lorry for twenty hours I think I would have done it for a Jungle breakfast but not, I must admit, in winter. There was no heating in those old cabs. The manufacturers hadn't even thought of closing the pedal vents with rubber screens, to keep the frigid air from screaming through, so I can only imagine what the Jungle Café meant to those drivers.

There were two 'transport caffs', as they were called, to give food and warmth to drivers going over Shap. The Eagle's Nest on the summit of Shap was the other, concocted out of two buses joined at the rear like copulating dragonflies. There's an old picture of it on the web. I never saw it myself, but I'm sure it was a very welcome refuge. The winter weather over Shap is fierce, amounting to blizzards at times. The long, slow climb to the top on the old A6 was interminable and the descent was even more treacherous, especially for older lorries. Snow and ice covered the surface, there were many sharp curves, and to add to the difficulties the road was often blanketed in thick fog. No doubt I romanticise the Jungle Café. If I were to see it today as it was then I would probably call it a dump, but such judgements taken out of the context of time and place have no real meaning. In the mid-forties, just before and after the war ended, there was nothing to compare it with.

I virtually never ate out. It would have been forbiddingly expensive, something done by posh people, and I knew of nowhere near me you could do it anyway, apart from British Restaurants. These were canteens set up by the government so people could eat where they worked, and they offered nothing better than school dinners. For a while my mother worked as a book-keeper for an advertising agent in Fleet Street, and once I took a bus to her office, though I can't think why. The agent, a Mr Greenwood, had a thing for my mum, apparently, and he slipped a pound note into my hand and told me to take her out to dinner at the Strand Corner House. She didn't care for him, and the bribe didn't work, but we did go to the restaurant, which, to my twelve-year-old eyes, was like a palace. That's the only time, I think, that I ever ate out in London.

Food, throughout my childhood and up to the time I went to Paris, was a primitive commodity put together in straitened circumstances. Of course I had likes and dislikes so I can say that my tastes were sufficiently well developed to recognise that the canteen lunches we were served at school were pretty awful. They consisted, as I remember them, of cabbage, meat, gravy and potatoes. The meat was often a very thin slice of over-roasted beef, but might also be Spam. The gravy was brown, watery and rather tasteless, and the meat swam in it with the potato. The cabbage was *Rocky Horror* green, having been boiled with soda. There were 'sweets', like spotted dick and custard, but all I remember about them was that they weren't very sweet. There was nothing much wrong with it as food. In fact, our wartime diet was healthier than anything working people had eaten before. At home we were rationed to very small amounts of meat, butter, cheese and sugar, all the really tasty stuff. Only vegetables were unrestricted, but the British had never really figured out what to do with them. Fortunately my mother was able to dig back to her German roots and took pains to make things appetising. I remember that she chopped up the spinach, the bane of every child's life, and mixed it with a white sauce that made it quite edible. The spinach back then was quite unlike today's light green, tasty leaves: it was dark and leathery and so packed with oxalic acid that it set your teeth on edge, which was why Popeye was pressed into service to sell it to us.

It was quite customary then to eat more or less the same

thing every day, just as we wore the same clothes every day. In fact, sameness was the outstanding feature of life in the forties. Every day was more or less the same, and Sunday was like the rest, but emptier. The interest and excitement in day-to-day life was what you conjured up for yourself, which I realise now was no bad thing. In my case it came from what I was learning in school, from books I was reading, from hobbies and, of course, the overarching excitement of a war in progress. What were Voroshilov's tanks doing today? How far had Monty got in North Africa? Food was only a minor source of pleasure. When I could get away with it I took the threepence I was given to pay for the school lunch and used it instead to buy a much tastier fishcake from a lady with a little shop in Fulham Road.

My eyes were opened to *haute cuisine* by a ship's cook, a friend of my mother's, who came over one day when she was sick and sliced up potatoes in front of my eyes at such speed that the knife, never more than a hair's breadth from his fingers, was a blur. I have never mastered that particular skill, but his expertise propelled me into cookery. Because I was often alone at home after school I did experiments with cooking, but the results were not inspiring. What I lacked was ingredients. Mayonnaise, they say, was invented under siege by the chef to Napoleon's Marshal Mahon, poor fellow, who had nothing to offer his master but eggs and olive oil. Well, we had neither. There was powdered egg and powdered milk and margarine, and I learned to make execrable omelettes, and scrambled eggs that rattled in the pan like grape-shot.

So, to come to the Jungle Café out of a black night on a lorry, into a warm room of scruffy men, with food frying and steam on the windows, and to sit down to a plate loaded with a full English breakfast – well, why would it matter what the place looked like? The food was marvellous. By contrast, I've been to a few modern motorway restaurants, Little Chef, for example, that for all their glossy graphics and hygienic bathrooms are as cheerless as they are sterile. I would expect to get more nourishment by eating the photographs of the food on the menu than from the food itself: a total triumph of style over substance, where even the style is in bad taste.

How on earth did we get from the Jungle Café to Little Chef? I'm sure Health and Safety had something to do with it, but it could be that the business model of the fast-food industry is most to blame. The art of eating well is reduced to the business of consumption. We had some great years in between, when the controls eased and vanished, when the fruit and exotic vegetables came in, and with them all the culinary styles of the Mediterranean and the Orient. I tasted my first avocado in Kensington in 1956. They hadn't even got to Paris five years earlier. Gradually, and with great excitement, the coffee houses opened, restaurants flourished, people learned about new and delicious foods, and the years of plenty rolled in.

Now there are wonderful meals to be found all over the UK, but they're expensive and we want to spend our money on gadgets and toys, so we've learned the fast-food tricks from America – we've discovered that with enough salt, sugar and fat you can smother the tastebuds for a third of the price. Back to George Orwell in 1935: '... there is the frightful debauchery of taste that has already been effected by a century of mechanisation ... take taste in its narrowest sense – the taste for decent food. In the highly mechanised countries, thanks to tinned food, cold storage, synthetic flavouring matters, etc., the palate is almost a dead organ.'

In a way I see the fast-food trap as a metaphor for our life today. You can load the salt, sugar and fat on to everything. Films? Think special effects, mayhem and superstars. Politics? Think non-stop news cycles, Murdoch and fear-mongering. Sport? Think grossly overblown multi-billion-pound circuses like Formula One and the Olympics. Lifestyles? Think second mortgages, everything on tick, drugs and booze *ad nauseam* (literally, at two a.m. on weekends). The excess is both wretched and mediocre. We know none of it's very good but we're too lazy to work for anything better. For many who grow up without any particular aim in life (an all-too-common condition today), the only thing that seems worth doing is to make money. Those who can prey on those who can't, but none of it is very satisfying. I am fairly sure that what really lies behind the latest rash of riots, which are happening as I write these words in 2011, is not criminality, or resentment, or original sin: it's boredom and a dim sense of self-loathing.

23

Not a Crappy Photographer

I found the A6 again just north of Kendal and, with most of the traffic on the motorway, the road has become a pleasure to ride. The soft green Cumbria landscape swelling and rising around me on a lovely July day brought with it a happy sense of wellbeing. I tried to find the shelf of land where the Jungle Café used to be but I must have missed it. No great loss. I knew there was nothing left. I'd seen photographs of the site, with caravans or mobile homes parked there, nothing of interest, so I went on, climbing slowly up the side of Shap. Even on such a pleasant day there were winding sections and sharp bends among rocky outcrops, wreathed in patches of quite heavy fog, enough to give me an inkling of what a winter crossing would be like. And then on the uplands I just gave myself over to the joy of moving through beautiful country, immersed in the scents wafted on the wind, one of the special pleasures that come with riding a bike.

The road took me through Penrith, then into the centre of Carlisle, on the border with Scotland, a big pedestrian area full of students and bicycles, and a pretty provincial railway station behind it, with arches and a clock tower. I was seduced by the artistic décor of a restaurant that claimed to be Mexican, offering ingenious items of fusion food, so I ate a border burrito, drank a cappuccino, and thought about my next leg to Dumfries and Thornhill. If I were a methodical chap, bound by his commitments, I would have continued up the road to Glasgow and found my way on to the Dumbarton road to seek out the cottage where the object of my adolescent infatuation had lived: sweet golden Helen, who launched at least two of my trips to the frozen north. What she thought of my persistence I will never know. She certainly didn't encourage it, but was always nice to me. It happens so often that boys fall for girls just a year or two older, and at any time later in life those years would have meant nothing, could even have been a bonus, but at the time of my adolescence they were an unbridgeable chasm.

At any rate, I lost no opportunity to gaze upon my beloved. I remember so clearly being dropped by my lorry driver at a bus stop in Glasgow, taking a bus along the Dumbarton road and, in my bone weariness, falling asleep on the bus. I must have slid sideways for I woke up in the arms of the motherly Glaswegian lady sitting next to me, and can still see the confused but indulgent look on her face.

I already knew, though, that to return there would be a hopeless and depressing quest. I had been back to Glasgow more recently, and of all the changes I had seen in Britain, Glasgow's were the most drastic. Even the infamous Gorbals had been transformed, and there was now nothing alongside the Clyde that even resembled the hovels I had once visited. Instead I continued to Dumfries, on my way to meet a 'rural photographer'. His email said:

> Do you ever come up to Scotland? I can't travel, but read travel books ... I'm just over the border in a wonderful valley called Nithsdale ... I'd like to take your portrait ... My neighbour, a retired shepherd, is a classic bike restorer ... we had to photograph the bike before it went back to the owner. The field had just had sheep in it and to get the low angle I'd had to crawl on the ground and yes you guessed it, I came back covered in sheep crap.
>
> Chris Frear,
> Rural photographer,
> Thornhill, Dumfriesshire.

I thought I needed to meet a man who couldn't travel yet could crawl around in sheep dung. At the same time, I thought, he was clearly a man of discrimination. If I were going to crawl around in dung of any sort, sheep dung would be my preference. It's been twenty-five years since I had sheep, but I remember them fondly, and those glistening burnt-umber pellets that they dispense freely around the pasture are nothing but organic treasure. Then there was the additional incentive that his website showed some really beautiful pictures of pastoral landscape. He had recommended the Thornhill Inn so I booked a room there and rumbled through Dumfriesshire to meet him.

Plainly this whole area I'd been travelling through lately is no stranger to water and I'd been lucky to stay dry. There were clouds above and moisture in the air as I rode on some sort of a ring around Dumfries. Too lazy to dig out the map I told my TomTom to take me to Thornhill and pretty soon what started as a decent little road turned rough and rugged. Then it began to rain. Not much. Not enough to force a decision. Just enough to make life miserable. The hedges closed in on me, wet leaves tried to slap me in the face, the tarmac reared up, puddles formed, the little front wheels skittered about on the bumps and, although I knew they were stable, I never got over the feeling that they were about to slide into the ditch. The road made sharp, blind turns but luckily there was almost no traffic. I'd forgotten that my TomTom was set for tiny roads, and when I remembered I was too far along to go back. Funny how difficult it can be to rise above expectations. I'd programmed myself to travel in comfort. In my normal travelling mode these conditions wouldn't even have raised an eyebrow. But I must admit that rain has always dulled my appreciation so, as I came up to Thornhill, I didn't absorb the beauty of my surroundings as I should have. And, even though I have no reason whatever to suppose that the Scottish Lowlands are particularly unsophisticated, I found myself thinking I'd be lucky if the place I was going to was half decent.

So the inn really surprised me. I found the unassuming entrance halfway down a block of houses on the main through street. There was a small hallway, with the pub door on the left and the inn on the right, but it was evident as soon as I entered that someone with exceptional taste had been at work. The interior was arranged with dark wooden furniture, unpretentious and obviously well made, and a bar counter faced the door, but it was the way the space was used that left such a strong impression on me. My room was nicely done, the menu was impressive, the people behaved with professional assurance, the inn was welcoming and orderly and, above all, calm. I was glad of that when Chris arrived for he is not the calmest of people. He came at about seven to join me for a meal, and before long he was frizzing me.

I had no idea what 'frizzing' meant, and even now I'm not sure, but every now and then he would interrupt the flow of information by saying, 'Now I'm frizzing you,' and after a while I came to understand that 'to be frizzed' must mean to be told more about any given subject than would be considered cool. Actually, I am fairly immune to frizz. I invite information and am usually very happy to have it dumped on me by the cartload. I absorbed happily everything he had to tell me about photography, about sheep and farming, and about Thornhill itself because Chris Frear has earned the right to be taken seriously.

His pictures of the rural world around him are more painterly than photographic. Beautifully composed in the classic manner, rich in all the subtle tones of earth and haze and vegetation, extraordinarily precise in detail and focus, they make me want to be part of the scene. Nothing would be more pleasing than to discover myself by chance in one of those photographs, as though I had somehow been leading a parallel life in this bucolic paradise. It was when I asked him to explain why he couldn't travel that I felt the frizzing sensation. He began to talk to me about his mother, whom he adores.

I have often mentioned my own mother in these pages. She played a crucial role, of course, in determining where and how I spent my childhood. She took immense pains to ensure that I got a decent education. More importantly, she burdened me with a pathological respect for the rights and feelings of others and I am constitutionally incapable of manipulating people. This has disqualified me for most corporate jobs, preserved my sanity, and forced me to lead an interesting life on my own terms. So, naturally, I revere my mother for liberating me, but I am critical of her too. Her Prussian work ethic was implacable. I wish she had allowed herself to have more enjoyment in her life. I wish she had smoked less, and drunk more, and fooled around a bit, even though it was probably on my account that she didn't. So for me she is a real, fully rounded person. If there had been a father in the picture, I would probably be talking about him too. As it is, I am happy to have had an interesting stepfather to stand in for him for a while.

It seems perfectly reasonable to bring one's parents, warts and all, into a conversation about life and its ups and downs. But Chris doesn't bring his mother into the conversation: she *is* the conversation. She suffers, it seems, from a mysterious ailment. She has been moving from one medical institution to another, being treated in a rather cavalier fashion with no apparent benefit, and returning to Thornhill in between. He cares for her and worries about her, and his life is inextricably involved in hers, which makes it impossible for him to travel.

'Come to the house tomorrow,' he said, before he left. 'I'd like to make a portrait of you.'

There was little to do in Thornhill that evening, but I was glad to walk up and down the main street of this small town, even though the rain was still coming at me in little spurts. Just by the way things were presented in the shops I thought I could detect a difference in aesthetic, a drier and more discriminating approach to things than I was used to in England. I was surprised by the variety of food, clothing and artifacts for sale, and passed the evening happily on my own.

Chris's living room, when I visited next day, had a good deal of paraphernalia related to his mother's care: a special chair, other bits and pieces. I learned that she was an excellent painter and writer, that she always put a brave face on the world, exceptional in every way. At one point he telephoned her at whichever medical facility she was in, reassuring her in a motherly fashion, offering her advice. Then he rearranged light fittings and furniture so that the illumination was exactly right for the picture he wanted to take of me.

I sat in a chair, a little oasis of order in a room full of chaos, while he moved my limbs around and, with infinite care, created the portrait he had in mind. I watched the picture on a screen as he showed me the small miracles he could perform on the digital image, managing the shades and colours, even lifting the corner of my mouth to create a different expression, an expression I didn't have but might have had. I found it fascinating, impressive and rather alarming.

Compared with the extraordinary degree of freedom I enjoy, Frear works within the tightest of constraints, yet he leads a fully productive, creative life. I found this very educational.

I left Thornhill in mid-morning for Stranraer, the Scottish port where a ferry would take me to Belfast, and I made no mistake this time. There were two ways to go. People at the hotel advised me to take the A75 back to Dumfries and on to

Stranraer. On the map it seemed like a very long way round when there was a red road that went straight through to St Johns, but I took their advice. Later, out of curiosity, I measured the two distances and they were about the same. I can only imagine the zigs and zags on the red road that the map must have glossed over.

Back on a good highway I was free to think more about what had transpired, and about mothers in general. It occurred to me, since I was on my way to an intensely religious country, that my mother's most valuable gift to me was to protect me from religion. She didn't rail against it, but she didn't expose me to it either. In our home it was simply left unmentioned, and I was allowed to discover for myself later what a lot of damaging nonsense is imposed on people in the name of God. I don't mind giving Him capital Gs and Hs. I don't believe in Him myself, but I certainly respect those who do. He is not the problem, but a number of young people I met earlier in my life were dreadfully screwed up by their conflicts with religion.

It is one of the odd things about religious belief that atheists are always expected to come up with proof, whereas it should, of course, be the other way round. I don't get into this much any more. When I argued at school and university I used to feel that I might actually have a chance of winning by persuasion, but later it became clear to me that belief is beyond reason.

There are many mysteries to be encountered on this earth. I have myself seen a ghost (or, as experts pointed out at the time, 'an astral projection') but do I really have to invent a god in order to explain it? At the age of eighty I feel entitled to hold myself up as evidence that it is possible to live a good life, have friends, influence people, be passionate about good things and do relatively little harm, all without His help. And if something frightful were to smite me down tomorrow it would not prove anything to the contrary. To those whose insolent bumper stickers say, 'If you don't believe in God, you'd better be right,' I say, 'If He's anything like you think He is, I've got nothing to worry about.' But the very idea that any of us mortals could know the first thing about God the Creator is so puerile that it reveals all religion as a sham.

Here endeth the rant!

24

The Deep Pockets of Stranraer

The road to Stranraer circles the coast of Wigtown Bay, where a huge row is blowing up over wind farms, then kisses the top of Luce Bay where they catch a kind of shark called tope that I'd never heard of but which seems to be guzzled regularly everywhere from Scotland to the Peloponnese. My ignorance never ceases to amaze.

Stranraer is a small town at the bottom end of a long inlet – about eight miles long, I'd say – stuck well out into the Irish Sea, and on the map it looks like a perfect jumping-off point for a sea voyage to Belfast. It's one of those places where the light and the wind and the underlying chill seem to keep it permanently scrubbed clean. Bracing, I think, is the word for it. I've only ever seen that word on the old railway posters for Skegness, but I should think Stranraer can probably outbrace even that blustery east-coast resort. I got there early because the map still deceives me – I can't seem to get the hang of the scale – but TomTom got me to the Fernlea Guest House in good order, and before long I was striding down Lewis Street to see the sight. I think there is only one sight, but I liked it very much. Very much indeed.

They call it a castle, St John's Castle, but it's the kind of castle your child would draw. It's a miniature medieval tower block, a four-storey high-rise. Each floor is about thirty feet by thirty, and it was built by some people called Adair in 1510. They lived, I believe, on the second floor above ground level in what was probably considered magnificence. Then it was passed on, through marriage I think, to the Kennedys, and then something peculiar happened. The Kennedys were given land and a barony, but they lost their house. The castle was taken away from them and used as a garrison to suppress the Covenanters, and if you were looking for examples of the murderous nonsense that people get caught up in through religion, you couldn't start at a better place. There were the usual thumbscrews and iron boots, of course, but drowning by tide was a new one for me. Poor Margaret Wilson was strapped to a stake out on the beach, and left to wait for the tide to come in and drown her. It must have been comforting for her to know that she deserved her fate, having been convinced of her 'total depravity' by the not-so-good John Calvin.

After that the castle became a gaol, mainly for nineteenth-century riffraff who floated over from Ireland to lift the Reverend's potatoes and cause other kinds of nuisance. They had a murderer too – say it with a Scottish accent and it sounds quite bloodcurdling – who was reprieved and sent to Australia, where he survived for another twenty-five years, apparently. His crime had nothing to do with faith: your chances definitely improved if you could stay away from religion.

Stranraer seemed like a perfect place to eat fish and chips (to tell the truth I can't think of a place that isn't) and there were several rival establishments to choose from. I chose the Central because it had a businesslike look about it, lots of buzz and animation, and because it had a history you could read about during frying time. A large part of my own history could also be told through fish and chips. It began in a tiny slot of a shop, tucked behind the cinema at Notting Hill Gate in 1943. Chips were a penny and a nice piece of cod was threepence, and they were all properly wrapped with salt and vinegar in yesterday's paper. I remained a loyal customer through the decades, as bombs fell, the premises expanded and the prices climbed into the stratosphere. The slum that surrounded it in the forties (my mother was once offered a mews cottage there for fifty pounds, but couldn't afford it) became bijou residences in the eighties, worth a million. When a piece of cod was a pound I dined there in the presence of celebrities (Roger Moore's voice once drowned all other conversations), and eventually I tried to establish my credentials as a founding customer, but they said, 'So what?' or words to that effect, and in retaliation I moved my custom up the road to Costa's.

I have received a similar brush-off from other favourite haunts. There's the Polidor in Paris's Latin Quarter, for example. I ate eggs with mayonnaise there in 1951, when I could find the money, and came back regularly as we all prospered, but when I tried to interest the *patronne* in my long-standing support of her restaurant she gave me a pamphlet about the people who'd eaten there thirty years earlier than me, including James Joyce, which effectively shut me up.

I can't believe I'm the only diner who has been humiliated in this way. We should form an association. Meanwhile, as I waited for my haddock, I read about the Petrucci family from Italy, who had opened the Central in the sixties. Not a very long history, admittedly, but they made the most of it. I had a very good meal, and went out to look around a bit more. I thought it best to make sure the ferry terminal was still there, since they were planning to move it, and then I went to Tesco's for a bottle of wine to take to my room. It's a good thing I was planning to pay for it. To my surprise, every bottle had an electronic security device clamped round the neck, which I took as a rather pointed comment on the population of Stranraer.

'Och,' said Kenny, at the guest house later, 'the people of Stranraer have deep pockets.'

It was foolish of me, I know, to expect my arrival in Belfast to be particularly exciting in any way. The ferry was like a hundred other ferries; the docks likewise were the usual mini-maze of one-way streets between walls and sheds. I had an address to go to, where a friend, Drew Millar, was expecting me, and TomTom got me through the traffic without difficulty and on to the Lisburn road, which seemed very similar to any number of other British roads. Even the confusion I experienced looking for his turning on to Windsor Park was no different from the kind of confusion I had experienced many times, and when I asked for directions, nobody said, 'You can't get there from here.'

Drew, a warm and fuzzy man, guided me in over the gravel drive past some very English box hedges.

'Bring on the leprechauns,' I cried. 'I've come all this way, and absolutely nothing Irish has happened.'

Very wisely, Drew said, 'Be careful what you wish for.'

'Well, everything looks the same as it did on the other side,' I said. 'The streets and the houses, I mean, although perhaps a little shabbier.'

'Actually they are the same,' he replied, 'with one exception that you obviously haven't noticed. All the windows in Ireland are set back into the wall.'

Amazing. I looked, and I saw that it was so, and I thought, What of it?

But I was just sulking. Actually, there were big differences, but they were more chronological than physical. I had arrived during one of Belfast's relatively rare periods of peace and quiet, in 2010. The Good Friday Agreement had held so far and those with an appetite for crisis could be satisfied by looking south at the financial débâcle that was ravaging Eire. Drew and his wife, Ruth, had found themselves a very nice house of many floors, instantly recognisable to me, but from London in the late fifties, my favourite period, which was also the age of renovation. That was when, in London, you could still find good solid old houses that had limped through the war on life support and been put to various temporary uses, such as bed-sits (or brothels, like the house opposite ours in North Kensington, run by the infamous Peter Rachman). There were solid pine floors that could be sanded down, anaglypta to be stripped off the walls, false ceilings that came away to reveal fine plaster mouldings, and so on. Belfast had simply been fifty years in aspic while the Troubles raged on, and was now catching up.

Drew and Ruth were halfway into the recovery process, so that I climbed the staircase on frayed and mismatched carpeting past glimpses of perfection as though through a museum of DIY. The garden at the back was a gem, and we were soon sitting there with wine and waffle, but still nothing Irish, except Drew's magnificent Belfast accent, often imitated, never equalled.

Ruth is assistant principal of a quite posh girls' school, and Drew got me to do a talk in one of the classrooms for a bunch of bikers he had assembled, but again there was nothing specially Irish about the occasion, and we even had dinner at a Chinese. But finally Drew succumbed to my persistent whining and took me to the Irish heart of Belfast, the Cathedral Quarter, which has recently been reclaimed from neglect. He'd heard that various notables of the press would be imbibing at a famous pub called the John Hewitt, and they certainly should have been, for it's a splendid place, but in fact they were round the corner on Hill Street at Nick's Warehouse. That was where I found myself across the table from a large, handsome woman, called Anne Hailes, who works for Ulster TV, and from that moment on the Irishness never stopped. She told me stories about her heroes, and I was reminded of how fortunate we were now to be able to walk around in this city without being blown up.

Her top hero was a pugnacious ex-rugby player called Jim McDowell who is the Northern editor of a rag entitled the *Sunday World*, which specialises in all the most scandalous stories of the week, among them, of course, the ones to do with naughty priests. It has always amused me that the juiciest religious scandals are reserved for the Lord's Day. The *Sunday*

Pictorial and the late unlamented *News of the World* were equally eager to sanctify their pages with religious news, and while I'm on the subject I need to right a grievous wrong. Somehow, during all the gossip about masterpieces of tabloid journalism that surrounded the Murdoch affair, the greatest of them was never mentioned, and that was Harry Procter's wonderful double-page screamer in the *Sunday Pic* when he exposed a delinquent reverend. It read: 'Go Unfrock Yourself Father Ingram'.

But these are not the stories that put McDowell into Anne Hailes's pantheon of great hacks. Jim McDowell is a truly fearless crusader against organised crime, in particular the crime that was organised by criminal remnants of the Troubles, and has put his own life at risk many times.

'One day he opened his front door and was shot point blank,' she said. 'He still has the bullet lodged in his spine, a few millimetres from his spinal cord. He knows who did it, but would never go to the police. He asked friends to take care of him.'

Martin O'Hagan, a *Sunday World* reporter, was shot dead by the LVF, a loyalist paramilitary group, in 2001, and while McDowell was attending a court hearing in 2009, involving men charged in connection with the murder, his car was vandalised. Less than two weeks later he was attacked and savagely beaten in the street only fifty feet from the front door of the Belfast City Hall. After a good repast of such lurid tales, washed down with Guinness, we sauntered out into the street where there appeared before our eyes, like a magic mushroom, a beautiful red taxi.

In large measure the next few pages are an account of the world according to Billy Scott.

25

A Mystery Tour

The magic red taxi belonged to Billy Scott who is 'licensed by the institute' to tell stories about Belfast, and having heard from Drew that I was a writer, he not only took us home but declared himself eager to come and fetch us next day. The written word is held in high regard in Belfast – perhaps because the spoken word is all but impenetrable, by me at any rate. Only if you have heard Belfast spoken would you understand.

So up rolled the little red cab in the morning to take us on a mystery tour of Belfast, city of bombs, bullets and poetry. Drew and Ruth sat comfortably at the back, and I balanced on the folding seat by the window with my recorder in my hand, so that I could catch the drift of the narrative, but first Billy wanted to know what kind of a Californian celebrity he had snared.

'A lot of sunshine?' he tendered, no doubt picturing the beaches of Malibu.

I had to disabuse him. 'Yes, sunshine, of course, but it's not at all like the California you're imagining. Just as you'd make distinctions between the North and the South, I live in northern California and it's completely different. You can't swim in the ocean. It's too cold. There are mountains and forests and bears. It's the Wild West, actually. The valley that I live in really was a Wild West valley, complete with a cattle baron who lived in a white mansion, and a smooth henchman who poisoned wells and drove the settlers out.'

Billy laughed. 'You know,' he said, 'I had a woman from California in the cab once. I asked her, "What's your impression of Ireland now?" and she said, "Your cows are dumb. They just stand there and eat, doing nothing. Our cows are alert. They look to see what's going on around them." And that was her overall impression of Ireland.'

We rumbled along the Lisburn Road and then on to the university.

'That's Queen's University on the right. It's one of the Russell group,' he said.

'What's that?' I asked, always suspicious of anything that sounded like corporations taking over. But Billy explained that a lot had happened since I'd left Imperial College sixty years ago. Apparently the twenty top universities in the UK met at a hotel in Russell Square and formed a cabal to hog all the research money for themselves. They got two-thirds of it. They call themselves the Russell Group.

'There's a statue to Ernest Walton,' said Billy pointing. 'He got the Nobel Prize for discovering the proton.'

Well, not quite, that was Ernest Rutherford, but Walton *was* Irish, and he did fire protons at stuff. Not that he ever worked in Belfast. He was in Dublin. I suppose the statue was wishful thinking. We passed the College of Theology, which had been run by the Presbyterian Church since 1851, and then came the Student Quarter where Billy had found a favourite place to eat.

'Their all-day breakfast is terrific – fried bacon, fried sausages, fried eggs, fried tomatoes, fried soda bread, fried potato bread, baked beans and chips.'

It's a canteen that should be on an arterial bypass. Billy is not a slim man. He's plump in a jovial kind of way and he lets his shirt hang out over his belt, but he is certainly not fat either, and I should really have complimented him on his restraint. The meal is a well-known recipe for obesity, called an Ulster Fry, and it's the Irish equivalent of the full English breakfast, only more so. Drew had already promised to expose me to it, as a rite of passage.

Billy's cab rumbled on to Shaftesbury Square, 'named after the Earl of Shaftesbury', and now Billy was in his element, dispensing juicy scandal, with suitable emphasis.

'The eleventh earl died recently, not long after inheriting the title from his father, the tenth earl, and the tenth earl was murdered in the South of France by his wife and her brother. Work that one out. The aristocracy! Setting examples to the rest of us.'

A fifty-foot-high white wall, belonging to the Ulster Bank, loomed up in front of us, and two dark figures, one above the other with legs outstretched, apparently floated against its surface far above us. They were put there by Elisabeth Frink (later a dame) in the sixties. She didn't give them a name, but the citizens of Belfast, who enjoy ridiculing art, have supplied one.

'The lower one,' said Billy, 'is called Draft. What d'ye think they call the one above it?'

We struggled briefly and failed.

'Overdraft, of course.' Groans and laughter.

'And over there's the restaurant, the Cayenne, owned by the famous chef Paul Rankin. He's the one with the big knives. But if you meet him in the street, don't worry. A recent United Nations report said that Belfast, after Tokyo, is the second safest city in the world – for tourists.'

It may be now, but it certainly wasn't then, as I was reminded when the Europa Hotel came into view.

'Anyone who came here to do business stayed at the Europa, so the IRA constantly bombed it. It was bombed twenty-eight times in the seventies and eighties,' said Billy, with relish. 'It was an economic target.

'And over there, on the right, is the Crown Liquor Saloon. It was owned by a Catholic and his Protestant wife. She insisted on calling it the Crown, so he had a mosaic made of the crown at the front door, so that everybody who enters walks on the crown . . .'

'Well,' I said, 'I can match that story,' building my reputation as the guide's passenger from hell.

Back when the *Daily Express* was a powerful newspaper owned by Lord Beaverbrook and selling four million copies, and I was a young reporter, two distinguished men stood at the commanding heights: the editor, Edward Pickering, and the gossip columnist, Donald Edgar. Both were tall, slim and expensively suited, and frequently went out together for lunch. The entrance hall of the building had a composition floor, and embedded right in the middle was a large circular design of Lord Beaverbrook's Crusader, St George, flourishing his sword and defending the British Empire. One day, from the stairs above, I happened to look down on the hats of the dapper duo as they sauntered towards the glass doors, deep in conversation and obviously destined to walk over the Crusader. I held my breath. Just at the periphery of the image, still talking, they swayed elegantly, one to the left, one to the right, walked round it, joined up again and left the building.

I understood then that everybody was afraid of the intemperate old baron.

We rolled on, past the opera house, '1895 ... *Indio Romanesque,*' chanted Billy, and then the Presbyterian Church headquarters.

'Presbyterians being Scottish are very canny with their money, so they leased the floor as a shopping mall and used their theological expertise to frame the terms and conditions of the credit-card agreement, against which many of us have greatly sinned.

'This is Wellington Place, named after a famous Irishman, the Iron Duke, the Duke of Wellington ...'

'He was *Irish?*' I exclaimed. Somehow that fact must always have escaped me, but Billy was quick to soothe my concern. The Irish had never thought of him as one of them either. A famous Irish orator was reported to have said, 'The poor old duke. To be sure he was born in Ireland, but being born in a stable does not make a man a horse.'

And then came Belfast City Hall, built in 1906, 'where the

council manages the affairs of the city of Belfast', which gave Billy the hook he needed to retell, with obvious enjoyment, the story of one of the arcane political intrigues Ulsterites seem to find so compelling, when the Alliance Party supported the Sinn Fein candidate to defeat the Democratic Unionist Party for deputy mayor, 'which upset the apple cart'. None of this meant anything to me, of course, but it had obviously provoked a storm of abuse at the time. The loser, Eric Smyth, was reported to have told the Alliance Party member, 'Your hands are covered in blood, you shameless traitor.'

'Now,' said Billy, 'see the blue plaque on the wall. This is dedicated to the man who revolutionised agriculture in the twentieth century, Harry Ferguson. He invented the tractor.

'And that hotel over there – *Cosmopolitan* called it one of the top ten sexiest hotels in the United Kingdom.

'And there's the Northern Bank, scene of the UK's and Ireland's biggest ever cash robbery. But of course it couldn't have been the IRA who robbed twenty-six and a half million pounds – don't be daft – because, according to the government, the IRA aren't involved in any criminal behaviour any more.'

The robbery was a remarkable operation. The robbers had seized the managers, taken their families hostage, then driven a van up the side-street, filled it with money, and even come back for a second helping. The police had managed to find a bit fairly soon.

'They recovered fifty thousand pounds hidden behind the toilets of the police athletic grounds. Cheeky.'

We passed an edifice with a row of busts above the windows.

'The first one is George Washington, but at the very end

here – you see the man with the turban? He's the Maharaja of Cooch-Behar, and he was the man who codified the rules of snooker. Came from Calcutta in 1888, the same year Belfast got its charter as a city.'

Then we were in Chichester Street.

'That was Arthur Chichester. In recognition of his military service during the nine-year war, James the First made him Lord Deputy of Ireland.'

Chichester had been utterly ruthless. He imposed draconian punishments left, right and centre, and squeezed the Irish nobility off their lands until eventually they left for Spain in the Flight of the Earls. Scottish Protestants were given the lands, 'which formed the model for the plantations of Virginia'.

'And the round building in front of you is the second-best conference centre in the world. It has hosted everything from the Three Tenors, to the World Council of Toilet Bowl Makers.'

My remark about Thomas Crapper got the treatment it deserved.

'Now here's the Customs House Square, on the left. This was originally Belfast's Speaker's Corner, where members of the public could climb up the steps to spout their politics before they started shooting them for doing it.

'You see the bronze pig on the steps? That's representative of the famous Belfast orator Mr Ballantyne. And one of Mr Ballantyne's famous speeches was where he gave out about the evils of ping-pong ... which could lead to even more depravity, like line-dancing.

'And the blue plaque is for Anthony Trollope who worked in Belfast as a civil servant. And the blue fish here is representative

of the regeneration of the waterfront area. Every scale ...'
somehow he made the word 'scale' stretch out like an eel
'... every scale on the fish tells a different story.'

Ten or so years ago, all of this large waterfront area alongside
the River Lagan had been derelict. It was hard for me to comprehend how much of Belfast, a city of half a million, had
been blighted by the bitter sectarian strife that lasted for
decades, making it impossible to attract money or enthusiasm
for improvements. Like many other uprisings, as with that of
the FARC in Colombia, for example, what began as an honourable revolt against oppression had long ago descended into
unbridled criminality, with violence simply begetting more violence. It's a wonder that it ever stopped. But here in Belfast it
has, and in this new era of peace the Laganside Corporation
has spent at least a billion pounds cleaning up the mess, creating 16,000 new jobs at the same time.

'There'll be a twenty-storey building here,' said Billy. 'Of
course, being from California, you couldn't comprehend a big
building like that ... but they'll give you a parachute once you
get above the third floor.'

A huge new arena stood by the river, the Odyssey, home of
the Belfast Giants, an ice-hockey team. They had proved to be
very successful.

'Currently they're British champions, but the arena can be
converted into a ten-thousand-seat concert hall as well. We've
had Lady GaGa, and Beyoncé ...

'They've cleaned up the river too. There's salmon and trout,
and now the otter have come back. We have to satisfy their
sophisticated palate – for salmon, trout and suckling pig. No,
not suckling pig, that's not really true.'

I spotted a transport museum and asked Billy what was in it.

'Oh, there's the world's first vertical take-off plane, that was built by Shorts ... lots of steam trains and that sort of thing. Then there's the *Mayfly*. The first woman in the UK to successfully build and fly her own plane was Lilian Bland. She lived out at Carnmoney. She called it the *Mayfly* – as in "may fly". She was the first woman in the UK to do that. It's never really been recognised.

'And this is the Harland & Wolff shipyard. It's where we built the *Titanic*. It was our claim to fame – that we built the unsinkable ship – but that's not in the Ulster Transport Museum. The *Titanic* was only one of over two thousand ships built here. Another ship built at the same time as the Olympic class liners was the SS *Nomadic*. She was a tender ship for the White Star Line.'

The more I heard about *Nomadic*, the more she surprised me. Here was a ship, 220 feet long, with luxury accommodation for the upper classes, a thousand people all told, and yet her sole purpose was to ferry passengers from the dock to the grand ocean liners standing out at sea. What huge expense for such a minimal role! She once carried Charlie Chaplin, Elizabeth Taylor and Richard Burton to one or other of those liners, and she ferried J. J. Astor to his fate on the *Titanic*.

Immediately I remembered the time I had drifted out on a dark night from Southampton to one of those immense ships, moored there in the Channel, dazzling me by its display of lights, like a floating castle. I was only six, joining it for the last leg of the liner's schedule from New York to Hamburg. I must have been on a tender like *Nomadic*, but sadly I remember nothing of it. Even the name of the liner has gone. I used to think it was *Deutschland*, but have since discovered that it couldn't have been. Even so, it gives me great pleasure to have been, however briefly, part of that vanished age.

Nomadic carried troops in both world wars, and continued tendering for Cunard White Star until 1968. For a while she was a restaurant ship, berthed in Paris. Then the restaurant closed, she fell into disrepair and was towed off to Le Havre to be broken up but was reprieved and rescued from the scrap-yard for £171,320 by the Northern Ireland government. Now, in 2010, a hundred years exactly since her keel was first laid, she is being restored as a museum ship for the Belfast water-front.

'That's her right there,' said Billy. 'She was built of the same materials and by the same people as built the *Titanic*. She's in a restoration programme, costing ten million pounds, the whole thing, toothpicks and all.

'Of course you've heard the great conspiracy theory? You see, the *Titanic* had a sister ship, the *Olympic*, and the story goes that she got damaged so they brought her in to port, switched the names, sent her out into the Atlantic and sank her for the insurance.'

Which is nearly the most stupid and gruesome conspiracy theory I've ever heard.

'Harland & Wolff was the biggest shipyard in the world, but over there was another smaller one. They called it "the wee yard" where they built refrigerated ships, and as a conse-quence, do you know what else was invented down here? Air-conditioning. Why anybody in Belfast would have wanted air-conditioning I've no bloody idea.'

We drove on to a wide open area of gravel in front of the museum, and could see that a number of vintage cars had been brought out on display. Drew became agitated, because he had spotted a blue Riley sports car, looking very like those old thirties racing cars, with straps holding down the barrel-like bonnet. This was apparently the car of his dreams, so we all got out to stretch our legs and have a look.

Then Drew fixed his car-lover's gaze on another – an MGB GT. 'Oh, he shouldn't be allowed in,' he said, with mock disgust. 'He's got the wrong wheels on. It's very sad, isn't it?'

But my eyes were riveted to a quite different kind of vehicle and at first I thought I was dreaming. There was the same black Vauxhall police car with the pointy nose that I had dragged out of my memory from those post-war hitch-hiking years, and flanking it were two very convincing-looking policemen, one a bobby with his helmet on, and the other an inspector with tie and peaked cap. It felt like a message, as though someone was saying, 'Good job. Keep at it.' Ridiculous.

The other thing that struck me immediately was how small the car was. Those two big men sitting in it must have been jammed together shoulder to shoulder, like an Ealing comedy. I couldn't help imagining Terry-Thomas and Peter Sellers pursuing Sid James, the rascally villain, and furiously ringing their bells.

Ah, yes, the bells. I asked one of them if the bells were working. Without a word he leaned in and the bells pealed merrily.

'Don't you sometimes get the urge, when you're driving around, to ring the bells?' asked Drew.

'Oh, yes, I do it all the time when I drive past the port police. They always give me a big grin.'

The car had been privately owned when they acquired it, but it came originally from Cheshire, and they told me how they had taken it back to England and found the retired police inspector who had once used it. He was now ninety-four years old, but he still had photographs and arrest reports from his active days, and the uniforms. He, too, found the car smaller than he had remembered. They had spent hours listening to his stories.

'It was only for me to drive around in,' he had told them. 'There was only me, two sergeants and ten constables, but I couldn't be seen riding a pushbike.'

We all climbed back into Billy's little red coach and headed across the bridge into town, passing a tall, ethereal sculpture, a figure of a woman holding a huge hoop over her head – ethereal because it was made of widely spaced thin iron rods swirling upwards, and really rather beautiful.

'She's the Lady of Lagan,' said Billy, 'fifty-seven feet high and supposed to symbolise peace, love and happiness. We call her the Thingy with the Ringy. And over there, on the hill, you should be able to make out a man's face – he's lying on his back, looking up at the heavens. This gave Jonathan Swift his inspiration for *Gulliver's Travels*.

'Of course if you believe that story you'd be right bloody gullible.' Probably because, as far as I can make out, Swift was not in Belfast for long and, in any case, had no need of a mountain to stimulate his imagination.

We came into the Cathedral Quarter, also totally neglected for decades, and Billy pointed to a building covered with columns. There was one row at the top of a flight of stone steps, and then another in front of the first floor, clearly intended to suggest stability of a high order. Now that you can buy a fibreglass column for a few hundred pounds, banks have had to give them up, since they aren't fooling anyone any more, but this was once the headquarters of the Ulster Bank. Now it has been rescued from dereliction and turned into a five-star hotel called the Merchant.

'See those vases on the parapet? Those are wine urns, and they were supposed to represent the institution's liquid assets. The first governor was checking his liquid assets too closely, tripped on the top step and was dead by the time he got to the bottom.'

But the restoration of this district is impressive, with clubs, pubs and restaurants opening everywhere. Belfast is not a place that's ever been on my list of missed opportunities before but Billy Scott has gone a long way to change my mind. What better time to move in when it's just coming to life after decades of torment? The place is drenched in history of the most contentious kind, with people having treated each other in the most cruel and heartless manner, usually in the name of God or some trumped-up authority, and yet it is clear to me already that these are kind, warm-hearted people willing to enjoy life.

Billy was in the middle of a dissertation on Presbyterianism. Apparently this was an area where Presbyterians came to settle in the seventeenth century, only to be hit by the Penal Laws in 1691, prohibiting all non-Anglicans from practising their religion, getting an education or holding public office, which put a pretty comprehensive dampener on life. If they wanted education they had to go to Dublin, where their heads were filled with heady notions of liberty and equality.

'Well, in 1791, in Peggy Barclay's tavern on Sugar House Entry a Presbyterian minister named William Drennan (who coined the phrase "Emerald Isle"), along with a Protestant lawyer from Dublin called Wolfe Tone and Henry Joy McCracken, founded a secret society down there called the Muddlers Club.'

My own parents must have visited Ireland some time in the fifties because I found a comprehensive and indigestible guide to Ireland put out by the Irish Tourist Board in 1951 among my mother's books. Amplifying Billy's remarks, it describes the wave of prosperity that infused Belfast through the eighteenth century, based mainly in the linen trade, which was developed through the expertise of exiled Huguenots. According to the guide, this was when Belfast was christened 'Athens of the North'. It must have been a heady time.

'Prominent citizens became patrons of learning and art, schools and libraries were established, and especially great interest was taken in the language, literature and music of Ireland.'

Apparently an outstanding cultural event of this period was the great Harp Festival of 1792, held in the Assembly Rooms of the Old Exchange, and that's something I would have liked to experience.

Of course it was all bound to end badly. Deluded into thinking that all this fine flowering gave them the moral strength to oppose their oppressors, like the Hungarians in 1956, the Czechs in 1968 and, no doubt, countless others I'm unaware of, they rushed to the barricades. Billy pronounced their doom succinctly.

'They rose up against the British in 1798 as the United

Irishmen and there were about thirty thousand killed in the uprising. McCracken was tried in the Assembly Rooms, and executed outside his own front door – just around the corner from where James Murray invented Milk of Magnesia.'

It would have taken oceans of the stuff, I suppose, to soothe the dyspeptic rancour that has plagued the Irish since the British planted their surrogates on Northern Ireland.

In search of relief ourselves, we all got out and walked to the Duke of York, down a narrow, cobbled lane that had been gorgeously decorated with flowers, red-painted benches and murals. There was even a weatherproofed piano or two standing outside. Billy, of course, was very well known here, and we all got into vigorous conversation.

Generously refreshed, and after a short rendition of 'Chopsticks', we set off again to see what every tourist in Belfast is obligated to see: the murals of those terrible years when the Republicans, the Loyalists and the British Army were caught up in a seemingly endless chain of murder and retribution. We were pleasantly delayed for a while by a marching band that came surging round the corner, drums, whistles and brass making a beautiful commotion, and then we found ourselves at last in the badlands of Belfast, the streets around the Falls Road and the Shankill Road, where the ordinary people lived, and where the grim reality of the Troubles hit you in the face from every angle.

Of course they have their fascination, those miles of high fences and gates, with their murals and graffiti. The warring factions lived so close together, with only a street separating them, that space had to be made, houses had to be destroyed, and the blank walls left at the end of a row are covered in messages of hate and defiance. I find them ugly and depressing. The artwork is brutal and primitive, with touches of infantile sentimentality. I can think only of all the lives that were perverted and ruined.

Billy said there were moves to bring down the 'peace lines', as those barriers are called, and to find other ways to separate immiscible neighbourhoods, like replacing the walls with shopping centres where the employees have to be equally Catholic and Protestant. 'Anybody employing eleven or more employees, your workforce has to be representative of the community, fifty per cent Catholic, fifty per cent Protestant. It's called positive discrimination.'

Standing there in the bleak open space between the two warring tribes, I can't imagine how those murals were painted. They certainly took time. Whoever was up there on ladders or scaffolding painting those threatening images must have been ripe fruit for a sniper's bullet. I suppose the murals of menacing masked men in black pointing their guns straight at you with deadly intent will be preserved. Personally I would photograph them and then obliterate them, but I'm nobody's loyalist.

Billy led us through the whole horrible history from both sides with humour and erudition, and it was more than two hours after we'd started that we finally rolled up to Drew's front door once more. There had never been any talk of money changing hands, but obviously two hours of an experienced guide's time, let alone the use of the taxi, would cost a pretty penny, and as we all exchanged more jokes and said our goodbyes, it slowly penetrated that Billy was simply about to drive off and disappear into the city. With some difficulty I was able to force some money on him to cover, at least, the cost of the fuel.

It's not often that I benefit from veneration for authors. The only other country I know where they are held in such high esteem is France, and I was foolish enough to leave France at just the moment when my name became known, but even there I can't imagine being taken on such a spontaneous free tour. Where I live now I'm merely tolerated, and occasionally humoured. It's nice to know that there's always Belfast.

Drew and I went for another ride next day to the north of the island. It was a lovely day (which is saying something in Ireland) and we drove through beautiful countryside to the top of Northern Ireland where Drew pointed out the geographical curiosity that the Republic of Ireland, which lies to the south, actually goes further north than Northern Ireland, from Cardonagh to Malin Head. In fact, I'd been struggling quite a lot with the geography and nomenclature of Ireland. I was under the impression that Ulster was an indigenous word for Northern Ireland but I was quite wrong. Ulster was there first and has nine counties, but only six are part of the United Kingdom of Britain and Northern Ireland. I also believed that Eire was a name for the self-governing Republic of Ireland, which is right or wrong depending on which dictionary you ask. Ireland is also peculiar in that the border between the two bits is more or less invisible, yet the British bit uses pounds and the other euros. However, in practice none of this causes any problems at all, at all – which is an expression much employed by comic Irish characters in movies, but which I never once heard used in Ireland. Not to mention 'Begorrah!'

Well, on our way back we came through Ballymoney

because it was inconceivable to Drew that I would leave Northern Ireland without paying my respects to Joey Dunlop, and of course I deferred to his wishes and kept to myself the embarrassing truth that, although I dimly recognised the name, I was by no means sure what he was famous for. Even as I write these words I blush with shame, but nevertheless that's how it was. As we entered Joey's bar, almost the first thing I saw was that very fast Honda motorcycle hanging from the ceiling, with a number 3 on the tail, so I had a pretty good idea of what it was Joey Dunlop did for a living. I also grasped that he was no longer doing it, due to death.

My relationship with death-defying sports is complicated. I don't do them mainly because I'm scared. On the other hand, I do other things that scare me, so fear can't be the only reason. I understand the need to take risks, but not for their own sake. I have enormous respect for the determination, the discipline and the focus that enable people like Joey to bring everything they have so close to the edge, again and again. I know a lot more about Joey than I did when I entered his bar. I know that he won the TT Formula One in five consecutive years, that he rode and won a phenomenal number of races, that he is considered one of a handful of the most admired racers of all time, and that he was a smashing bloke who cared for his family, who did a tremendous amount for children in Romanian orphanages, and was honoured for it. And this is what gets me. He must have loved life as much as I do. He avoided all the nihilistic sectarian strife, must have seen it for the dreadful waste it is. With the mental and physical qualities he possessed, with his skill and dexterity, he would surely have succeeded at so many things. Why put your life on the line every time? Well, I've had this discussion with others, and there's a lot I obviously don't understand.

I felt the warmth of Joey's bar. Drew persuaded his widow, Linda, to come out and have her picture taken with me, and I'm glad I have that picture now because I have a better appreciation of what an extraordinary experience it must have been to be married to such a man. Then things took an odd turn. There was another man sipping his ale on a bench, and as Drew was introducing me to Linda, he got to his feet. He had come there, of course, to pay homage to Joey and now, to his surprise, he found himself with someone else he had always hoped to meet. There were more photographs, and I felt a little like an impostor. I hope Joey would have understood.

The next day had been chosen for my initiation into the saturated mysteries of the Ulster Fry. We visited St George's Market, a great market over by Donegal Quay. Drew and I joined the queue at the counter behind a bevy of well-upholstered Irish biddies and he showed me no mercy. I had eggs, bacon, sausage, baked beans, black pudding, fried tomatoes, mushrooms, potato bread, fried soda bread and, I do believe, several other ingredients lost under the steaming heap. The eating arrangements in a small roped-off area were a shade crowded. Two of the ladies allowed me to force my aluminium chair into a gap between them at a rather unsteady little table. Leaning back from their own massive platefuls, they murmured sweet words of encouragement as I bent manfully to my task. Not ten yards away a banner, raised over a soup tureen, glared down at me with all the force of a religious pronouncement: 'The Healthy Option,' it declared. I ignored it and consigned my liver to hell.

When it comes to judging between the full English and the Ulster Fry I have to say that the English version is sweeter, but the Ulster one is fatter, so I therefore deemed it meeter to polish off the latter. And after that I staggered away to inspect the stalls, particularly the fish stalls, which were rich in varieties I hadn't seen very often, like the conger eels I first came across in Chile.

Distracted by that long and terrible story of violence in the streets of Belfast, I took a while to realise just how important the sea and the building of ships has been to the city. From early beginnings in the seventeenth century, the industry grew to the point where about 25,000 were employed until the Depression of the thirties brought that figure down by almost 90 per cent. The Second World War pulled the manpower back, reaching 35,000, according to one report, and this lasted until the sixties before tailing off. In fact, my host, Drew, had worked in the offices of Harland & Wolff at one time.

I was also slow to notice that the Wolff in Harland & Wolff has two fs. This was not in pursuit of aristocratic pretensions, such as those sported by Colonel Ffarington Eckersley, but because Gustav Wilhelm Wolff was a German from Hamburg. Half of my family also came from Hamburg so I was quite used to hearing how close they felt to Britain, closer in some ways than to south Germany. They have a dialect in that northern part of Germany, called *Plattdeutsch*, which is so close to English that when the British occupying forces moved into the Hamburg area at the end of the last war they had no difficulty in conversing with the locals. And because Hamburg was so open to the world, through its maritime traditions, it was among the last to succumb to Hitler's venomous xenophobia.

237

It was quite natural for seamen and merchants to move back and forth across the North Sea.

Following the story of Gustav Wolff and his uncle Gustav Schwabe, you find a wonderful example of the tangled web that finance spun around the globe 150 years before the word 'globalisation' was coined. Schwabe was the son of a Jewish merchant and clearly a very clever man. Already at the age of twenty-one he had become the partner of an Englishman called Boustead who had started a business in Singapore, a business that still exists to this day. He married well, into an English family, and became a junior partner in a Liverpool shipping company called Bibby, another firm that still flourishes today. He and his nephew both moved to Liverpool, and the two Gustavs, between them, helped Edward Harland to acquire a small shipyard out on the mouth of the Lagan, which became Harland & Wolff. Naturally the first ships they built were for the Bibby Line, but Schwabe also knew Thomas Henry Ismay, who ran the White Star Line, and when Ismay expressed an interest in the Asiatic Steam Navigation Company, they set off east together on a long tour to the Orient.

Somewhere, over a famous game of billiards, Schwabe agreed to finance the White Star Line, which was bankrupt, provided that its ships were built by Harland & Wolff, and that was the beginning of a long line of famous ships emerging from what became the biggest ship-building yard in the world. I was amazed to discover that the last of those ships, *Olympic*, *Britannic* and *Titanic*, each carried as many as three thousand passengers across the Atlantic, so *Nomadic*, ferrying its thousand people out to the waiting liners, was only carrying the *crème de la crème.* How the other two thousand got on board I don't know. All in all, it is estimated that the White Star Line took as many as two million emigrants from various parts of Europe to the New World, many Irish among them.

But the prize nugget from my researches was the news that the fifth generation Bibby, Sir Derek, who was terminally ill with leukaemia at the age of eighty, committed suicide by swallowing aluminium phosphide. As a result his corpse began emitting lethal fumes and the part of the hospital where he was taken had to be shut down.

Drew had arranged for me to stay with a friend of his south of the border. I left him and Ruth on Saturday morning and rode out along the Lisburn Road until I got on the A1 south towards Dundalk where it changed to the N1, which was the only apparent difference between the North and the South. Then I went off to Ardee, to Slane, to Navan, then towards Dunboyne, and, following my TomTom, I eventually found myself at a crossroads with one pub and a couple of houses, called Nurney. The people hanging around it seemed content, and I suppose the scene was the closest I ever got to my imaginary picture of Ireland. Certainly the Celtic Tiger, which was supposed to have transformed the place, was nowhere to be seen. I preserved the illusion as long as I could. I had a drawing of sorts that should have told me where Drew's friend lived, but whichever way I turned it around it didn't seem to lead me anywhere, and I felt the tiger draining out of my tank too.

It seemed to me that if there was only one pub and two houses surrounded by fields and trees they would know if a biker lived nearby but I was wrong. They sold me a packet of crisps and half a pint, and then I went for a little ride around, and by pure chance, behind some bushes, I found a complete suburban development of detached houses with pastel stucco walls, wrought-iron gates and garages, straight out of a California realtor's brochure. The contrast was so dramatic that it was unreal. And, of course, the bubble had to burst because Drew's friend was called Jochen, which is not by any manner of means an Irish name.

A neighbour let me into Jochen's house, where Jochen found me a few hours later and carried me off to a bar of rather better character in Kildare, and there we bemoaned the financial melt-down, which had left Ireland teetering somewhere between Eldorado and the poorhouse.

Jochen works for a company in Dublin that installs security alarms but he himself is big and strong enough to frighten off any intruder, and he rides an appropriately big motorcycle. He was kind enough to suggest that we go off together next day to inspect the Wicklow Mountains, even though riding beside my tiddly three-wheeler might expose him to scorn and derision.

We rode south and crossed the Liffey at Newbridge, where I forced him to take picturesque shots of me by the river, and then into the national park from the north, through Blessington. There is a severe beauty about this landscape, though the mountains are really majestic, sweeping hills, largely treeless and darkly pigmented, clothed in mauve-flowering gorse and bearing the marks of generations who gathered peat from this soggy land. Between the hills clear streams run over peaty stone beds, children play, earnest hikers assemble and that day Kevin's Cones was whipping out the ice cream.

We came down eventually to a resort area by a lake, where a

fine restaurant beckoned to us with trestle tables out in the sun. Jochen told me it was the Loch of Glenmacnass.

'This is where Dubliners flock on a hot Sunday,' he said. Actually I believe they flock there on cold Sundays too, because they seem unable to tell the difference. Although the sun was out I found it quite chilly and kept my jacket on, but all around me a profusion of bare arms, legs and belly buttons was on display. I have to say that I have never seen people so willing to expose their flesh to gales and frost as I did in Ireland, and while I was cowering most of the time inside my sweater and jacket, my companions were still in their shirt sleeves and would have stripped to the buff if it had been deemed seemly.

This might give the impression that the Irish are a healthy people, bursting with vigour and loaded with longevity, but against that it must be said that I have rarely heard so many people discussing their bypasses, angioplasties and stents. Somewhere, no doubt, there will be figures to support or demolish this idea, but my impression is that the Irish heart carries a curse on it. What a shame, with all that lovely butter and cheese.

We came out of the park through Rathdrum, where we were held up by a wheelbarrow race being run through the centre of town. Everything considered, it was a fine day and a lovely ride, and if there was anything wrong with it at all, at all, it was the continuing absence of Irishness, except in the names of the towns (and, of course, the recessed windows), which leads me to think that I have been badly misled over the years by a fantasy foisted on me by Hollywood and the Irish Tourist Board, and reinforced by wishful thinking.

*

241

I was off to Dublin next day, but something went wrong. I'm not sure what it was. There was a fellow I was expecting to meet who never materialised, and maybe I was putting too much stock in that. The weather broke up and started drizzling. I got myself into a hotel on O'Connell Street, but it was the wrong hotel to be in, as bland and unIrish as you could get. Up the road was the dead writers' museum, in a very fine eighteenth-century building where I was happily sheltered from the miserable weather. I boned up on my knowledge of Shaw and Joyce and Yeats (knowing this wasn't what I should really be doing) until I was ushered out.

Drew had given me the names of a couple of pubs that sounded promising. I crossed the River Liffey and made my way through the drizzle to Poolbeg Street and Mulligan's pub, reputed by some to have the best Guinness in Dublin, and where it was said living writers were sometimes to be found. I having failed to give them the joyous news of my arrival, they had all stayed at home. The only notable event that did occur was in the Gents, a rustic affair with a galvanised iron trough running along the wall. A large, florid man, expensively dressed in loud dog's-tooth check trousers, was on his mobile phone: he was obviously a criminal engaged in some terrible plot while I stood by helpless to intervene.

'I'm telling you,' he shouted menacingly, 'you've got to find her or you know what'll happen. Do ye hear? I want that girl found tonight.' Then he stared at me, and I crept out.

Shamed by my inability to coax anything much out of one of the world's most romantic cities, I slunk off south the following day. I had to console myself with the thought that at least I'd done Belfast proud. I had a date with a ferry sailing out of

Rosslare the next morning and a bed in a B&B close to the port, so I rode down the coast from Dublin to Wexford, hoping that somewhere along the way I might get entangled with somebody or something, but a cloak of invisibility seemed to have fallen over me.

I might have visited the James Joyce Tower, for in Ireland they *did* give a novelist a tower. It wasn't built for him, but rather to repel Napoleon, but I was done with dead writers, and in any case I was too far along before I'd even thought of it. I had considered visiting Bray, and looked it up on the Internet, but the travel guide told me not to bother: 'Its grand old buildings are now somewhat shabby, and the town itself is not very attractive ... the next town on the coast, Greystones ... is much nicer.' The top comment on Greystones read, 'The guy in Iretons isn't that friendly. He chased me across the park cuz he thought I was robbin biscuits.'

Wicklow is a dormitory for Dublin and had obviously fed off the Celtic Tiger phenomenon. It has a nice wide harbour (in fact, all the harbours I saw on this coast looked great) and, given a day or two, I can imagine enjoying it. Why its Irish name means 'Church of the Toothless One' is still a mystery to me. I went on, doing my best to stay off the motorway, and tried to follow the coastline.

My last day in Ireland was a catalogue of broken dreams. All the way down this coast, with its little ports, and cliffs, and beaches, obviously made for summer holidays, there were ghost towns, developments like the one Jochen lived in, but they were either unfinished or empty. I saw signs boasting of delights that never came to fruition. In Kilmuckridge, a name to conjure with, I was promised a 'European Menu: Launch June

2009', only to be brought down to earth by the Kilmuckridge Centre of Learning where courses in child-minding and gardening are available (and why not, indeed?).

And then, before I knew it, I was in Wexford, another fine port, and, whatever its destiny, this one at least has a grip on its past. For Wexford has a hero, John Barry, born of a poor Wexford farmer in 1745, who created a dramatic career for himself in the US Navy; in fact, he practically created that navy. As a commodore, he was the first to capture a British warship, and he fought the last naval battle of the American Revolution. There is a fine statue of him in Philadelphia, but the one I saw in Wexford is brilliant: he stares eager-faced out to sea with his cape flapping behind him in the breeze and his hand on his sword.

I found a little pub on the quayside with wooden floors where I had a fairly miserable meal, being too late for the lunch special, and wandered around the town a bit but saw nothing of great interest. By this time I was feeling that I hadn't really seen Ireland at all. One day I shall return and do the whole country in a little red taxi, but meanwhile I must live with my disgrace.

Epilogue

I finished my wanderings with a rather new appreciation of this sceptred isle. While Britain appeared, from a distance, to have become a radically different country since my youth, I am now more inclined to believe that the changes do not go very deep, while there have been real changes in me. Viewed from America, the British scene has seemed like an endless parade of politicians, bankers and bureaucrats, crises and statistics, apathy, greed and occasional outbreaks of violence and drunkenness. Now all that seems like superficial nonsense, and I think I have found an underlying structure that has remained intact for centuries and could hardly be ruptured or dispelled in the foreseeable future. My fears were mainly to do with overcrowding and controls, and it is certainly more difficult for an individual to carve out a space for peace and tranquillity these days, but I've seen it done, and am immensely reassured.

Much of the alienation I felt was my own invention. I moved a long time ago to a remote and backward village in the South of France and found a degree of freedom from controls and surveillance that was positively idyllic, but it depended very much on my willingness to live a very simple, bare-bones existence. Somehow I convinced myself that this would not have been possible in the UK, but I know now this was an unfair judgement. By the same token, my move to a piece of land in the mountains of northern California gave me extraordinary liberty to build what I liked and how I liked, but again at the cost of being quite distant from society. It was a choice I was happy to make, and while I don't for a moment regret any of those decisions, I see that there were other choices I could have made that would have led to a happy outcome in England.

There is something about the English temperament that is still very appealing, and when I find myself among a group of people I am always most comfortable if they are British. Trying to define that temperament defeats me. It is easier to say what it is not. Coming from California, I find that the British don't smile much, but I don't believe, as some Americans do, that this is because they have bad teeth. I think it's because they don't like to advertise. They prefer to stay private where Americans like to be public, and I like that about the British (and also, I must say, the French). In a sense, being American is all about doing business, and it's difficult to get away from the feeling that you're being sold something, even if it's only a sunny personality or a perfect set of teeth.

Going back now to the beginning, before we all became consumers, I see that my life could have gone in any number of ways. In those first post-war years when I was beginning to explore the world the future seemed bright and exciting. Almost every day great discoveries were announced promising a new world of comfort, health and freedom from work. Nuclear power, it was said, would provide unlimited energy at virtually no cost, revolutionising our lives, with never a mention of the possible dangers. I can remember reading in the dear-departed *News Chronicle*, my mother's paper at the time, how entire houses and their contents would be constructed of clean, cheap, labour-saving plastics, with dust-gathering vacuum systems built in. The wonders of penicillin were justly celebrated, and new antibiotics were queuing up to rid us of every known disease. The Science Museum demonstrated such wonders as tubes that conducted light, machines that could think, and a strange putty-like substance that flowed and bounced, though nobody knew what to do with it. Obviously a golden age was glowing just around the corner and the big question was, What would we do with all our leisure time?

Much more important to me, though, than any of these scientific advances, was the birth (or, rather, rebirth) of the coffee-house. I had become used to the drab meanness of the war years so every small improvement was a delight. The first one I remember was at Notting Hill Gate, between two cinemas, where some unsung genius managed to acquire a Gaggia coffee machine from Italy, and install it in a small shop with chairs and tables. Finally, for the price of a cup of coffee, there was somewhere congenial to go and meet other young people – meaning, of course, girls. But the best of these temples of free thought was the Troubadour on Brompton Road. The 31 bus came back into service, carrying me to the birth of

a new age; because it was in the coffee-houses that the real revolution of the sixties germinated.

I found the Troubadour through a Pole called Freddi, though
how I found Freddi I can't remember. In a sense Freddi, or
someone like him, has been with me throughout my life. He was
one of those rapscallion figures who was almost too disgraceful
to be seen with, yet too ebullient, charming and exciting to dismiss. He was invariably unshaven (in a way that would be
fashionable today), his clothes were stained and crumpled, and
he always looked dirty. Yet somehow he managed to hold down
the job of secretary to a rather formal golf club in Watford. He
drank at least one bottle of whisky a day and chain-smoked – he
must have been in the black market to sustain his habit – but his
stories and humour were irresistible. He tried to get me involved
in a business he had invented for providing ready-made sandwiches to office workers, a brilliant idea at the time, but the
place he had found to assemble these treats was disgusting and
his fingernails were filthy. I have never been keen to get next to
godliness but he was too much even for my anarchic views on
hygiene. My last fond memory of him is at the Troubadour
when he came in one evening with one of those curly wooden
hat-racks on his head and called out, 'Look, ever' body. I'm a
Wiking.' The point of this story is that everybody knew him,
and laughed, and for the most part we all knew each other, and
there was a sense of community and sympathy between us,
which was like the beginning of something wonderful.

What I really want to do is start this journey all over again.
Now that I have discovered TomTom's brilliant ability to guide
me along that fine tracery of unnamed, unregulated, ancient
farm roads, I want to follow them from one end of Britain to

the other. Then I want to buy a narrow-boat and spend another year just drifting. I was afraid that progress and population had filled Britain, but I have a better grasp now of the scale of it. There's a huge amount of empty country left. It's the cities and the roads that give the illusion of a totally congested nation, and jobs, of course, force people into these crowded hubs and corridors.

Work is the curse of the drifting classes, as gypsies discovered long ago. If you can do without a job, a place of work, you don't have to rush with the lemmings. The fact that most people can't do without a job – and many actually enjoy their jobs – certainly works in my favour, just as it does where I live now. California could even be regarded as a wildly over-stretched version of Britain, with the vast majority of the population huddled into a relatively small space in southern California, while the north is virtually empty. My home is in the empty bit.

I couldn't count the number of times I have been asked to explain why I still choose to live in this remote and, in some ways, primitive valley. Of course, there are many reasons, and I vary my answers to relieve my boredom. It is mid-winter as I write this, and I look out on a brilliant day of sun and blue skies and glorious scenery – what visitors have called 'a million-dollar view' – but there are great views everywhere, and that's not really why I'm here. It's because I am pretty much at liberty to do as I choose on my own land. It is true that even that freedom is being slowly chipped away, and I have to be cautious, lower my profile, even resort to a little harmless subterfuge – but, then, wherever you live the price of freedom is eternal vigilance, and a little cunning.